Embodied Carbon for Sustainable Building Conservation

This timely volume provides the latest research, guidance, examples, and methods for understanding, calculating, leveraging, and reducing embodied carbon in building conservation. In the context of climate change and increasing energy costs, imperatives to replace or substantially modify older and historic buildings are rapidly accelerating. The idea that a new or replacement building will perform better overlooks the embodied carbon of that which it replaces. In effect, the pressures of one conservation agenda, that of energy efficiency, threaten to eclipse another, that of heritage. The embodied carbon of existing buildings must be addressed if calculations of operational energy use are to be properly balanced.

In this book, an international and multi-disciplinary group of authors offer perspectives on the influence and implementation of strategies to account for embodied carbon for the conservation of the historic environment. Examples are deliberately diverse and extend beyond buildings to the valorisation of a heritage grassland landscape specifically because of its capacity to store carbon, to the fundamental attributes of historic concrete and our responsibility to consider replacement with critical care.

This book inspires confidence in developing arguments by spreading examples globally and delivering plausible, actual narratives alongside clear up-to-date guidance. It brings together international standard-setters with practitioners, academics and advocates, all clearly explained. It also illustrates how embodied carbon played a pivotal role in seeking to determine the case of saving Marks and Spencer, Oxford Street, London, from replacement. This will be an essential resource for all building conservation and heritage practitioners including building surveyors, architects, conservators, engineers, conservation officers, building archaeologists and consultants.

Oriel Prizeman is Professor of Sustainable Building Conservation at the Welsh School of Architecture, Cardiff University. Formerly a practicing architect, she founded an MSc at Cardiff (2013) and Centre (2021). She was a board member of the Association for Preservation Technology (USA) 2015–2017 and of Architects Accredited in Building Conservation (UK) since 2024.

Embodied Carbon for Sustainable Building Conservation

Edited by

Oriel Prizeman

Routledge
Taylor & Francis Group

LONDON AND NEW YORK

Designed cover image: Demolition of BBC Broadcasting House, Llandaff by Dale Owen of Sir Percy Thomas & Sons 1967, 4th November 2021 © Oriel Prizeman

First published 2025
by Routledge
4 Park Square, Milton Park, Abingdon, Oxon OX14 4RN

and by Routledge
605 Third Avenue, New York, NY 10158

Routledge is an imprint of the Taylor & Francis Group, an informa business

© 2025 selection and editorial matter, Oriel Prizeman; individual chapters, the contributors

British Library Cataloguing-in-Publication Data
A catalogue record for this book is available from the British Library

Library of Congress Cataloging-in-Publication Data
Names: Prizeman, Oriel editor
Title: Embodied carbon for sustainable building conservation / edited by Oriel Prizeman.
Description: Abingdon, Oxon ; New York, NY : Routledge, 2025. | Includes bibliographical references and index. |
Identifiers: LCCN 2024061512 | ISBN 9781032864099 hbk | ISBN 9781032864037 pbk | ISBN 9781003527404 ebk
Subjects: LCSH: Sustainable construction | Buildings--Retrofitting--Case studies | Historic buildings--Conservation and restoration--Environmental aspects | Historic sites--Conservation and restoration--Environmental aspects | Urban renewal--Environmental aspects | Value analysis (Cost control) | Carbon dioxide mitigation | Carbon sequestration
Classification: LCC TH880 .E45 2025 | DDC 363.6/90286--dcundefined
LC record available at https://lccn.loc.gov/2024061512

ISBN: 978-1-032-86409-9 (hbk)
ISBN: 978-1-032-86403-7 (pbk)
ISBN: 978-1-003-52740-4 (ebk)

DOI: 10.4324/9781003527404

Typeset in Corbel
by SPi Technologies India Pvt Ltd (Straive)

Contents

Figures

Tables

Contributors

Deepankar Kumar Ashish PhD, specialises in sustainable construction materials, focusing on low-carbon cement-based systems. As a researcher at the Faculty for the Built Environment, University of Malta, he advances the durability, reactivity and environmental performance of cement. His work supports global efforts towards net-zero construction and achieving sustainable development goals.

Henrietta Billings MRTPI FRSA is Director of SAVE Britain's Heritage, a national building conservation charity, a chartered town planner with a background in journalism, editor of *The Brutalist London Map* and *Art Deco Map*, and co-editor of *Tbilisi: Preserving a Historic City*. She is a trustee of the Royal Greenwich Heritage Trust.

Mahdi Boughanmi holds a Bachelor's degree in architecture from IUAV University of Venice (2013) and a Master's degree in Technological Innovation and Design of Urban Systems from the same institution, graduating in July 2016. His expertise extends to CAD/BIM technical drawing and 3D rendering with a specialism in HBIM.

Peter Cox FRSA is founder of Carrig Conservation. He was past president of ICOMOS' International Scientific Committee on Energy, Sustainability & Climate Change (2013–2020). He represented ICOMOS on the CEN Expert Technical Committee that produced European standard EN 16883:2017 "Conservation of Cultural Heritage – Guidelines" for improving the energy performance of historic buildings.

Jigna Desai is a Professor at the CEPT University, Ahmedabad, India and heads the Center for Heritage Conservation at CEPT Research and Development Foundation. She is an architect and works with questions of transformation, preservation and conservation of historic built environment and heritage places.

Riccardo Maddalena PhD, is Associate Professor at Cardiff University's School of Engineering and directs the DURALAB, an advanced accelerated ageing testing facility. His research targets concrete industry decarbonisation by repurposing mineral waste and byproducts. With expertise in cement and concrete science at nano- and bulk scales, he leads innovative projects on supplementary cementitious materials and life cycle assessments.

Mihoko Muto PhD is a heritage consultant. Since joining Japan Cultural Heritage Consultancy in 2013, she has worked on various conservation schemes for architectural heritage. Her research focuses on the conservation of cultural landscapes and sustainable development of industrial communities related to Tatara, the traditional Japanese iron-making method.

Camilla Pezzica PhD is an academic with a background in digital methods in architecture and urban studies. She is interested in inter-disciplinary research bridging disaster risk reduction and sustainable development, with a focus on spatial modelling and scenario-based approaches. Her contributions include models and tools for emergency management, heritage conservation and town planning.

Oriel Prizeman PhD is Professor of Sustainable Building Conservation at the Welsh School of Architecture, Cardiff University. Formerly a practicing architect, she founded an MSc at Cardiff (2013) and Centre (2021). She was board member of the Association for Preservation Technology (USA) 2015–2017 and of Architects Accredited in Building Conservation (UK) 2024–.

Rajan Rawal is a Professor at CEPT University, Ahmedabad, India. He works on building and city energy performance, urban heat islands, energy, thermal comfort, building codes and building material characterisation. His work involves field studies, laboratory experiments and modelling simulation.

Ahmed K. Taher PhD is a lecturer at the College of Engineering, Architecture and Environmental Design of the Arab Academy for Science and Technology & Maritime Transport. His research focuses on principles of sustainable conservation, aiming to preserve the built heritage environment for future generations, while his professional pursuits extend to private architecture practice.

Yoshitaka Takahashi PhD is president of the Japan Sogen Network and the chair of Aso Grassland Restoration Committee, Japan. He is specialised in grassland ecology and grassland management. Since joining the Ministry of Agriculture, Forestry and Fisheries in 1979, he has worked on a wide range of grassland-related issues.

Preface

The futility of fighting long-term problems with short-term political interests is nowhere more pressing than in respect to climate change. The urgency cannot be overestimated and for those of us who have, or aspire to have, some agency with respect to the built environment, the notion of stewardship and/or adaptation has already begun to overtake that of free agent of innovation in design. However, whilst this responsibility is understood by many, there is still an urgent need to augment the powers of persuasion on the framing of arguments in support of this mindset. The aim of this new volume is to loosen conventional boundaries of academic discipline and practice as well as global reach. With disparate evidence drawn from practitioners, academics and advocates based in Ahmedabad, India, Alexandria, Egypt, Cardiff, Wales, Dublin, Ireland, London, UK, Msida, Malta and Tokyo, Japan a shared conclusion is demonstrating its resilience.

Acknowledgements

With thanks to all those who have contributed and their supporting networks as well, of course, to my ever-patient family.

For Chapter 2: Peter Cox would like to acknowledge a number of colleagues who have been instrumental in our work in this area over the last number of years. They are Professor Aidan Duffy PhD, Aneta Nerguti, Ruth Graterol and Andrew Lundberg.

For Chapter 4: The authors would like to acknowledge inputs from Nigar Shaikh for material characterisation studies and Sneha Asrani for embodied energy calculations.

For Chapter 5: Funding: This work was supported by the Arts and Humanities Research Council "Shelf Life: Re-imagining the future of Carnegie Public Libraries" project [grant number AH/P002587/1]. This book is derived in part from an article "Carnegie libraries of Britain: Assets or liabilities? Managing altering agendas of energy efficiency for early 20th century heritage" published in *Public Library Quarterly* 01.11.20, Copyright Oriel Prizeman, available online: https://doi.org/10.1080/01616846.2020.1826242

For Chapter 6: This has been adapted and translated from Japanese for the first time from a paper entitled: *"Agricultural landscape around Mt. Aso and its ecosystem services: from landscape valuation by means of the cultural landscape concept through landscape rehabilitation to world cultural heritage"* (2017), published as an outcome of a Japanese National funding scheme.

For Chapter 8: This first case study was undertaken by a team at Carrig Conservation including Peter Cox, Dr Caroline Engel Purcell and Sinead Hughes. The case study that follows is taken from a report prepared by the Research arm of Carrig for the Government of Ireland (2024) by Peter Cox, Dr Aidan Duffy and Aneta Nerguti: "Energy Upgrading of Traditional Buildings for Low Embodied and Life Cycle Emissions Guidance and Case Study 2024".

1

INTRODUCTION

Oriel Prizeman

Since the publication of "Sustainable Building Conservation; Theory and Practice of Responsive Design in the Heritage Environment" (Prizeman 2015), 10 years ago, there has been a transformation (by no means causal, simply correlated) in attitudes towards existing buildings. The 1970s term "adaptive reuse" has become mainstream in architecture, distinguished from, but conscious of and importantly not entirely at odds with, that of "conservation" (Lanz and Pendlebury 2022). That said, there are much deeper roots with parallel sympathies that have been evident in architectural education for a significant period of time. The long-held philosophical championing of *continuity* was first exposed by Dalibor Vesely and Mohsen Mostafavi in an exhibition of Diploma Unit 1 at the Architectural Association in 1982 (Vesely 1982; Vidler 1982). These principles underpin core values to be expounded in schools of architecture at the influential Universities of Cambridge and Harvard for subsequent generations.

Rationale for the Book

Essentially the concept of embodied carbon is a means to account for the deficit owed to justify deconstruction or demolition. Professor Rhaman Azari and Alice Moncaster's recent edited book "The Routledge Handbook of Embodied Carbon in the Built Environment" (Azari and Moncaster) provides a comprehensive overview of the application of embodied carbon to the consideration of design for the built environment as a whole. At Carlton University in Canada, the definition of "Heritage Waste" was identified by Susan Ross and colleagues (S.M. Ross 2020; S. Ross and Angel 2020), explicitly tying these more marginal activities to the purview of conservation practice for the first time. In terms of sustainable building conservation, the focus has previously been limited to historic buildings as a relatively niche area of fixed or even diminishing scale. However, as the examples here cumulatively illustrate, with embodied carbon, it is frequently the relatively modern buildings, often unlovable or as yet unrecognised, that tend to emerge as the most challenging examples.

"Retrofitting" has become a commonplace and sometimes controversial shorthand for addressing energy upgrades to existing buildings

DOI: 10.4324/9781003527404-1

over new build. In 2019, the UK based Architects' Journal championed their "Retro-first" campaign proposing: "a major reduction in the consumption of raw materials and energy in the built environment through the adoption of circular economy principles. It opposes unnecessary and wasteful demolition of buildings and promotes low-carbon retrofit as the default option" (Hurst 2019).

In 2022 the then president of the Royal Institute of British Architects (RIBA), Simon Allford, launched its first "Reinvention Award" saying

> Looking ahead to the low carbon future, it is vital we always consider how we can reinvent existing buildings to work even better when they accommodate new uses. The careful husbandry of existing resources - including buildings – has a long and noble, if recently forgotten, architectural history that we are relearning – and fast.
>
> (Allford 2022)

The prize is now annual and evidently the theme is now truly mainstream as the current RIBA President, Muyiwa Oki, attested at the awarding of the prize in 2024

> The importance of retrofitting and adaptive re-use cannot be overstated. Not least because most buildings that we will inhabit in the future have already been built. As architects, we are faced with the task of creatively responding to this issue, while balancing the needs of the local community and environment,
>
> (Oki 2024)

echoing the prophetic words of preservation architect, advocate and former AIA president, Carl Elefante: "the greenest building is .... one that is already built" (Elefante 2007) almost two decades previously.

The Opposition of Heritage and Architecture in the 20th Century

In many ways, the late 20th-century divergence between modernist and conservationist architects was not by design, rather by an overly narrow interpretation of recent thinking implemented through rebuilding programmes since the Second World War. Misgivings towards the concept of architectural interventions unfolding on what Walter Benjamin referred to in the 1930s as a "tabula rasa" (Aureli 2013) and applied to post-war, post-industrial mindsets are closely associated with Le Corbusier, perhaps most dramatically in his proposal to completely erase the Ethiopian capital of Addis Ababa in a masterplan commissioned by Mussolini (Zeleke 2010). With respect to embodied carbon, this specific issue is critically important to acknowledge as this 20th-century modernising mindset is now viewed as an opposing manifesto.

Writing first in a series of essays in the journal *L'Esprit Nouveau* (Le Corbusier 1920) just over 100 years ago, the French modernist architect,

Charles-Édouard Jeanneret, known as *Le Corbusier*, had challenged the aesthetic values attributed towards engineered and new as opposed to architectural and canonical heritage. He presented two photographs and asked:

> Let us display, then, the Parthenon and the motor-car so that it may be clear that it is a question of two products of selection in different fields, one of which has reached its climax and the other is evolving. That ennobles the automobile. And what then? Well, then it remains to use the motor-car as a challenge to our houses and our great buildings. It is here that we come to a dead stop. 'Rien ne va plus.' Here we have no Parthenons.

He juxtaposed conventional aesthetic assumptions of beauty and utility withdrawing distinctions in order to reveal new values which were to transform architectural design in the latter part of the century. Indeed, writing in 1960, Reyner Banham described Le Corbusier's widely acclaimed 1923 polemic later published as a book, "Vers Une Architecture / Towards a New Architecture" (Le Corbusier 1923) as "One of the most influential, widely read, and least understood of all the architectural writings of the twentieth century" (Banham 1994 [1960]).

Global Focus

Whereas the scope of Le Corbusier's influence in terms of international declarations relating to modern architecture is well known, his simultaneous involvement with the Conservation movement is less frequently acknowledged – perhaps a further over-simplification. He himself had in fact helped to draft the first charter of the International Council on Monuments and Sites, the Athens Charter of 1931 (ICOMOS 1931). Confusingly, two years later he also co-authored another significantly influential Athens Charter of 1933 for the fourth meeting of the Congrès internationaux d'architecture moderne (CIAM). This included a section on "the legacy of history" which noted the need to protect "fine architecture" but also that "68. If their present location obstructs development, radical measures may be called for, such as altering major circulation routes or even shifting existing central districts – something usually considered impossible" and "69. The demolition of slums surrounding historic monuments provides an opportunity to create new open spaces" (Congres Internationaux d'Architecture Moderne (CIAM) 1933). While it is clear that Le Corbusier qualified his own comments towards heritage, the global scale of influence of his polemic has been read simply.

The intentions of the inter-war manifestos of ICOMOS and CIAM related to their architect's imperatives in a period of significant geopolitical instability. Le Corbusier's prescient focus on the motor car, anticipated and accurately emphasised the unprecedented growth to come. Indeed, an American annual report of 1920 proudly noted that "if all the motor cars and trucks in the United States were placed in a row four feet apart they would encircle the globe" (National Automobile Chamber of Commerce 1920). Global automotive numbers were deemed to have risen from in the range of 2m in 1920 to over 95m by 2016 (Qualman 2019). Today, architects stand amongst all who scramble to redress the impacts of that potential unfurled as a climate crisis that is another unprecedented global concern. The United Nations' Sustainable Development Goals aim to underpin agreed pathways. The Intergovernmental Panel on Climate Change highlights the disproportionate debt and thus duty to future generations (IPCC 2023). This global agenda again transcends a regional or national approach, hence the relevance of the geographical spread of examples presented in this book.

Husbandry

More nuanced understandings of curation and "husbandry" (Allford 2022) are becoming mainstream in architectural discourse as a sub-sector of sustainability. It is evident that the amplification of re-use, retrofit, retro-first and creation in architectural endeavour are not necessarily mutually exclusive approaches. Agents and stakeholders associated with buildings and sites have varied interests, needs and vulnerabilities. It is important to balance these as far as is possible. For the purposes of Sustainable Building Conservation, the concept of embodied carbon simply presents an opportunity to derive some empirical quantifiable evidence to determine the impacts of waste in an effort to balance the projected gains of greater energy performance.

Design of the Book

This volume deliberately draws together very diverse narratives in terms of professional role, approach, academic discipline and socio-economic context. It presents a handful of deeply considered case studies addressed by practitioners, academics and advocates in order to demonstrate that these issues have an intentionally global reach. With the intention of probing the concept of long-term husbandry or stewardship from a perspective beyond that of building conservation, a narrative of the carbon sequestration achieved through traditional grassland practices in Japan is included. The aim is to demonstrate the breadth of the question and the scope of the potential application of these principles for designing our future to manage and adapt rather than rebuild. The beneficial impact of acknowledging embodied carbon is illustrated across policies and practices of management, husbandry and design intervention at the scale of typologies, landscapes and buildings across continents and climates. The book is designed in three parts, essentially 1. Definitions, 2. Extrapolations, 3. Applications and each may be read independently.

Part 1: Definitions

Part 1 forms a reference resource for the reader who is seeking the capacity to frame an argument relating to the importance of acknowledging embodied carbon. Definitions are drawn from the scale of the building to that of the chemical and physical composition of materials. The relevant principal conventions of accounting for embodied carbon building conservation policy and practice are set out. These are underpinned further by a definition in terms of material science. The contributors in this section include Peter Cox, ICOMOS International – Past President of International Scientific Committee on Energy, Sustainability & Climate Change (2013–2020) and founder of Carrig International, a conservation research and sustainability practice based in Dublin, Ireland.

A review of embodied carbon from the perspective of materials science is presented by Dr Ricardo Maddalena who leads the unique DURALAB materials testing facility at Cardiff University, Wales together with Dr Deepankar Ashish, a researcher at the Faculty for the Built Environment, University of Malta. This chapter draws attention to the carbon intensive significance of concrete as a material.

Part 2: Extrapolations

In Part 2, three contributors demonstrate how the principle can be applied to a condition at scale in three very different contexts.

In India, Professors Jigna Desai, head of the Center for Heritage Conservation at CEPT Research and Development Foundation and Professor Rajan Rawal, an expert in energy efficiency, systematically review the importance of considering embodied carbon for modern heritage there. This critical and thorough review highlights the burgeoning urgency of acknowledging the legacy of 20th-century carbon intensive concrete buildings – a global responsibility and a difficult inheritance shared by many.

I present the findings of research carried out for the AHRC-funded Shelf-Life Project from an article originally published in *Public Libraries Quarterly*. The notion of a typology or identification of a pattern demonstrates how efficiencies in assumptions of energy use for library buildings built at the same time at the turn of the century can be used. These principles are derived from over 400 extant early 20th-century buildings in the UK. They can directly be applied to a further 2100 related Carnegie Library buildings in the United States and extrapolated further to the many public buildings using the same building technology of the golden age immediately preceding the First World War across the UK.

Expanding the notion of husbandry to the scale of landscape, an exemplar of traditional practice management of grassland is outlined by Dr Yoshitaka Takahashi PhD, president of the National Grassland

Restoration Network and the chair of Aso Grassland Restoration Council, Japan and Dr Mihoko Muto PhD, a heritage consultant. The issue of stewardship is illustrated here through the traditional expectation of practises for grassland management. This chapter takes a consideration of embodied carbon beyond the realm of buildings. It highlights the importance of acknowledging anthropogenic influence and responsibility to the cultivated environment more broadly.

Part 3: Applications

Finally, in Part 3, a sequence of deliberately diverse case studies have been selected in order to extrapolate the profitable application of considering the debt to future generations through sustainable building conservation acknowledging embodied carbon. In Alexandria, Egypt, Dr Ahmed Taher, lecturer in the College of Engineering, Architecture and Environmental Design department at the Arab Academy for Science and Technology & Maritime Transport, outlines the challenging rehabilitation of a typical 19th-century apartment building. Outcomes are systematically modelled using computational fluid dynamics to demonstrate the potential for a revision to passive strategies for ventilation to avoid the reliance on air conditioning in a warm and humid climate.

A case study of a Georgian house in Dublin, initially presented as a report to the Government of Ireland, is used as an authoritative and repeatable exemplar supplied by Peter Cox. In addition, the approach designed towards a highly significant student accommodation block in Trinity College Dublin is recounted.

Finally, from the perspective of advocacy, an incisive narrative of groundbreaking tactics is made by the UK's foremost anti-waste heritage NGO founded in 1975, Save Britain's Heritage. The 2024 SAVE campaign to challenge the decision of the Secretary of State on the future of an early 20th-century central London department store demonstrates how the message of re-use can garner national interest and potentially shift political decisions. With a background in planning and journalism, Henrietta Billings' detailed account demonstrates the leverage of perseverance and the breadth of support that media coverage can garner to deliver the real impacts of these calculations.

Reference List

Allford, Simon (2022). RIBA launches Reinvention Award to celebrate reuse of buildings. https://www.architecture.com/knowledge-and-resources/knowledge-landing-page/riba-launches-reinvention-award-to-celebrate-reuse-of-buildings?srsltid=AfmBOo0jkWYyHslERi4zPyE3qWpGB4THq_H8hGao3Rer5OD1ys7lmiNp. Accessed 11.11.24.

Aureli, Pier Vittorio (2013). The theology Of Tabula Rasa: Walter Benjamin and architecture in the age of precarity. *Log*, *27*, 111–127. http://www.jstor.org/stable/41765790. Accessed 11.11.24.

Azari, Rahman, and Alice Moncaster. n.d. *The Routledge handbook of embodied carbon in the built environment*. Routledge.

Banham, Reyner (1994 [1960]). *Theory and design in the first machine age*. Oxford: Butterworth Architecture.

Congres Internationaux d'Architecture Moderne (CIAM) (1933). The Athens Charter. https://designmanifestos.org/congres-internationaux-darchitecture-moderne-ciam-the-athens-charter/. Accessed 11.11.24.

Corbusier, Le (1920). L'esprit nouveau: revue internationale d'estetique, edited by Ozenfant, Amédée Dermée. Paul, Le Corbusier: Editions de l'esprit nouveau.

Corbusier, Le (1923). *Vers une architecture. Esprit nouveau*. Paris: G. Crès.

Elefante, C. (2007). The greenest building is … one that is already built. *Forum Journal*, 21(4), 26–38.

Hurst, Will (2019). Introducing RetroFirst: A new AJ campaign championing reuse in the built environment. *Architects' Journal*. https://www.architectsjournal.co.uk/news/introducing-retrofirst-a-new-aj-campaign-championing-reuse-in-the-built-environment#:~:text=The%20AJ's%20RetroFirst%20campaign%20proposes,retrofit%20as%20the%20default%20option. Accessed 11.11.24.

ICOMOS (1931). The Athens Charter for the Restoration of Historic Monuments – 1931. Adopted at the First International Congress of Architects and Technicians of Historic Monuments, Athens 1931. https://www.icomos.org/en/167-the-athens-charter-for-the-restoration-of-historic-monuments. Accessed 11.11.24.

IPCC (2023). Summary for Policymakers. In: *Climate Change 2023: Synthesis Report. A Report of the Intergovernmental Panel on Climate Change. Contribution of Working Groups I, II and III to the Sixth Assessment Report of the Intergovernmental Panel on Climate Change* (Geneva, Switzerland). https://www.ipcc.ch/report/ar6/syr/downloads/report/IPCC_AR6_SYR_SPM.pdf. Accessed 11.11.24.

Lanz, Francesca, and John Pendlebury (2022). Adaptive reuse: A critical review. *The Journal of Architecture*, 27(2–3), 441–462. https://doi.org/10.1080/13602365.2022.2105381.

National Automobile Chamber of Commerce (1920). Facts and figures of the automobile industry. 14 v. https://catalog.hathitrust.org/Record/100114589. Accessed 11.11.24.

Oki, Muyiwa (2024). RIBA announces shortlist for Reinvention Award 2024: Transformative designs showcase the potential of retrofitting. RIBA. https://www.architecture.com/knowledge-and-resources/knowledge-landing-page/riba-announces-shortlist-for-reinvention-award-2024-transformative-designs-showcase-the-potential-o#:~:text=Collectively%20and%20separately%2C%20these%20projects,re%2Duse%20cannot%20be%20overstated. Accessed 11.11.24.

Prizeman, Oriel (2015). *Sustainable building conservation: Theory and practice of responsive design in the heritage environment*. London: RIBA Publishing.

Qualman, Darrin (2019). *Civilization critical: Energy, food, nature, and the future*. Fernwood Publishing. https://www.perlego.com/book/3534726/civilization-critical-energy-food-nature-and-the-future-pdf. Accessed 11.11.24.

Ross, Susan, and Victoria Angel (2020). Heritage and waste: Introduction. *Journal of Cultural Heritage Management and Sustainable Development*, 10(1), 1–5. https://doi.org/10.1108/JCHMSD-02-2020-116. https://doi.org/10.1108/JCHMSD-02-2020-116.

Ross, Susan M. (2020). Re-evaluating heritage waste: Sustaining material values through deconstruction and reuse. *The Historic Environment: Policy & Practice*, 11(2–3), 382–408. https://doi.org/10.1080/17567505.2020.1723259

Vesely, Dalibor (ed.) (1982). *Architecture and continuity: Kentish Town projects 1978-1981, Themes / Architectural Association; 1*. London: Architectural Association.

Vidler, Anthony (1982). Architecture and continuity. *AA Files*, 1(2), 74–79.

Zeleke, Elleni Centime (2010). Addis Ababa as modernist ruin. *Callaloo*, 33(1), 117–135. http://www.jstor.org/stable/40732799.

Part 1

Definitions

2

WHOLE LIFE CYCLE ANALYSIS AS AN IMPERATIVE FOR SUSTAINABLE BUILDING CONSERVATION

Peter Cox

I would like to acknowledge a number of my colleagues who have been instrumental in our work in this area over the last number of years, they are:

Professor Aidan Duffy PhD, Aneta Nerguti, Ruth Graterol and Andrew Lundberg.

Introduction

Buildings are currently responsible for 40% of EU energy consumption and 36% of energy-related greenhouse gas emissions (European Commission, 2021). As a result, the EU Renovation Wave initiative which aims to double the annual energy renovation rates in the next 8 years, was developed to focus on the circular economy of the built environment across Europe. The refurbishing and reuse of existing structures, rather than demolishing and rebuilding, is consistent with circular economy principles such as "design for reuse and recovery" (Adams and Hobbs, 2017) and "reuse the existing asset" (UK GBC, 2019).

In 2022, The Irish Green Building Council (IGBC) released the third draft of the *Whole Life Carbon in Construction and the Built Environment in Ireland* which projects that while residential operational emissions will likely reduce by 2030, the built environment emissions overall are likely to increase if the current rate of embodied carbon emissions growth is maintained. This highlights the importance of considering the whole life carbon of a building project rather than just its operational carbon.

The *Energy Performance of Buildings Directive* (EPBD) has been recently updated to include objectives recognising the importance of embodied carbon1 in the reuse of buildings in addition to their operational carbon emissions (Directive (EU) 2024/1275). Minimising waste and developing a circular economy in construction is a key policy objective for Ireland, with the Climate Action Plan (Department of the

DOI: 10.4324/9781003527404-3

Environment, Climate and Communications, 2024) setting out an objective to decarbonise the built environment through energy retrofits and to reduce embodied carbon in construction methods and materials, in line with the Renovation Wave and the EPBD.

Traditional Buildings

It is becoming known across the world that a large percentage of buildings, certainly pre-1945, can be classed as "Traditional Buildings" which also includes Heritage Buildings. The majority of these buildings have a solid wall or single leaf construction – they are mostly built from locally sourced building materials such as mud, stone, brick or timber. The materials used were most likely open pore materials which allowed moisture to pass in and out of the wall. This system then translated into more modern construction using concrete and as late as the 1960s we were still constructing buildings with single leaf construction until the first major oil crisis of the early 1970s.

Cavity wall construction then became popular – some with and some without insulation – and it wasn't really until the 1990s that thermal performance of our buildings became a major concern.

Heritage Buildings

Heritage Buildings tend to be older and usually reference pre-1919 buildings and this class or typology can also carry a listing as further protection, however many from this category are not listed and therefore are vulnerable to incorrect interventions being made in the name of "Energy Efficiency or Sustainability" this often results in "Maladaptation". Maladaptation is when a building is energy retrofitted without consideration for the base performance of that building, using incorrect calculative analysis and using incompatible materials such as non-porous insulations and not introducing managed ventilation. All of these will result in sealing a building, entrapping moisture within the building fabric and increasing the carbon footprint when the aim was to reduce the carbon footprint. This maladaptation can also result in introducing mould which can cause health issues for the occupants.

Consideration

One must always know, understand and consider the building that one is dealing with whether it is a single cottage through to a Georgian or Victorian building be it institutional building or a large State building you should always use a methodology and diligent process to know the existing thermal performance of the building before considering interventions. The core data that one should start with are

• Age and construction of the building
• Condition of the building
• If there is a heritage value to the building

- Building materials used
- Building construction
- Occupancy/use of the building
- Correct U Value of the walls, windows and roof of the building
- Construction of the floor.

Only then can a professional advise on the best strategy to prepare for an energy retrofit of that particular building. Once this information is known the professional can then devise and/or design a high performing energy efficiency strategy that will be best for the building, best for the occupants and best for the environment.

Thermal Analysis and Calculations

Knowing the base thermal performance of the building is essential and we find many of the tools offered today for calculating U Values, Interstitial Moisture Movement, Cold Bridge Analysis and other calculative methods to be inaccurate with defaults or standards that mean you are starting off with the wrong base information which will result in getting inaccurate information out that can have a detrimental effect on the existing fabric of an older building. Calculative output is only as good as the knowledgeable input to a calculation tool – if the building is misunderstood and the wrong information inputted then the wrong results will be delivered and used.

EPC (Energy Performance Certificate)

The EPC, or BER in some countries, is a good example – this is a tick box situation primarily designed for post-1974 buildings where the assessor may not be familiar with a building type, and all "solid wall construction" is given a default value of 2.1/2.2 U Value and this, through research and *in-situ* U Value Testing, is now proving misleading with some solid walls performing even below 1. If a wall that measures 1 in real time is given a default of 2.1/2.2 then the solution is to use increased insulation which is not required, can be damaging to the original building fabric and results in increasing the carbon in the retrofit when it is again really not required. Over insulating a building can also lead to potential health issues for the occupants by encouraging mould growth, damp and other unhealthy micro-organisms.

Disharmony Between Standards and Building Regulations

In many jurisdictions around the world there are significant differences between building regulations and building standards which does cause confusion and can lead to misinterpretation. Within Europe there has been much advancement in encouraging research in this field mainly led by Horizon 2020 and now Horizon Europe with realistic funding available for worthwhile research into the energy efficiency of our new builds and, more recently, encouraging research into the energy upgrade of our existing buildings. Individual State

Parties such as Ireland and Scotland have funded such projects that have produced excellent guidance in this area.

American architect Carl Elefante is responsible for coining the phrase "The Greenest Building is the one that is already built" (Elefante, 2007) (a recent study commissioned by Historic England entitled "Understanding Carbon in the Built Environment" proved this in finding the demolition of a simple dwelling over 60 years old and replacing it with a typical modern new build of the same shape and volume, even to NZEB standards, will take 63 years to pay back the lost embodied carbon in the original structure, multiply this for larger buildings such as early industrial buildings, early office buildings and apartment blocks, and we are heading for a major carbon crisis (Duffy, 2019).

Near Zero Energy Buildings (NZEBs)

Whilst this concept has been instrumental in the changing attitude of the general public, building owners and construction industry it is also now understood that it does not go far enough to make a real difference in the energy performance and, more importantly, the reduction of CO_2 emissions from construction in general. As an industry we should be grateful for this initiative but we should also be aware of its limitations and push for understanding knowledge and education on how the construction industry creates a very high volume of the worlds CO_2 emissions; concrete, steel and glass for instance being the highest offenders in the production of carbon emissions during manufacture and the transport of these materials. Our industry as a whole should be considering carbon at design level and not when a building has been built – we should always consider low carbon solutions.

NZEB was always calculated and based on "Operational Energy or Carbon" and totally ignored or disregarded "Embodied Carbon" and the carbon intensity in producing modern building materials, their packaging, their transportation, their waste factor and their potential recyclability at end of life. This brings in the idea of "Whole Life Cycle Carbon Analysis" which is what we, as an industry, should now be striving and driving towards. It's no longer good enough to say let's race to energy retrofit 500,000 homes a year to 2035 or even 2050. If we are only doing this retrofit to NZEB standards we will fail to deliver the level of emission reduction, and we may actually increase our overall carbon footprint whilst jeopardising our older buildings' fabric and introducing new health risks to their occupants.

Equally we must think about our new buildings and, as everybody knows, we have a housing crisis in most western countries and again governments are reacting to this pressure and saying they want to build hundreds and thousands of new buildings without thinking of the global risk to our environment, whilst other departments within governments are trying to implement climate action plans to save the planet and all these new buildings, yes they may be energy efficient from an operational energy point of view but we ignore the cost of the carbon intensive methods of modern construction materials that are common today.

Pause for New Thinking

If the construction industry wants to make a difference and contribute fully to mitigation against our climate crisis than it is time to pause and engage in reassessing what we are building, how we are building, what is the real environmental cost of constructing new buildings and energy retrofitting our older building stock, and develop a mentality of "Whole Life Cycle Analysis" that takes a "Cradle to Cradle" approach so that our buildings will contribute greatly and affectively to truly reducing CO_2 emissions across the world and that we will still have plenty of homes which will be efficient and truly sustainable in the long term. The industry owes this to the world and the (hopefully) generations to follow.

What is Life Cycle Analysis (LCA)

LCA involves assessing the total environmental impacts of a product or service over its entire life: manufacture, construction, transportation, operation, maintenance and end of life. Impacts may include resource depletion and various emissions to air, water and soil. Whole life carbon assessment (WLCA) considers life cycle greenhouse-gas emissions only and reflects the critical importance of climate change to the global environment and human society. This concept focuses on these carbon emissions, rather than other environmental impacts.

The general LCA process is formalised in the ISO 14040 and 14044 standards which describe the main steps that should be undertaken in any life cycle study. These steps are summarised below:

Goal and Scope Definition defines the purpose, motivation and objectives of the study, and describes the methodology to be used. As well as highlighting the study's motivation and intended audience, this step must clearly define and bind the system being studied, how impacts are allocated (e.g. whether impacts are allocated to by-products), the relevant impact categories (e.g. Global Warming Potential) as well as data requirements and assumptions. The "functional unit" must also be defined so that meaningful results can be compared with other studies providing the same function. The exact methodology used is chosen by the LCA practitioner but must be stated and must follow the ISO guidelines (ISO 14044, 2006). While the ISO 14040 9 can be applied to any product or service, EN 15978 is specifically tailored to assessing buildings. The EU's Level(s) methodology, which is based on EN 15978, addresses six environmental indicators, one of which is life cycle greenhouse gas (GHG) emissions.

Life cycle inventory (LCI) analysis involves the collection of data and selection of calculation procedures to estimate the flow of fuels, materials and emissions into and out of the building over its life cycle.

This must be completed both for all embodied and operational stages and involves data collection, analysis and validation.

Life cycle impact assessment (LCIA) involves estimating the environmental impacts of the system based on the inventory data compiled and calculated above. It includes the collection of results for the different impact categories chosen in the Goal and Scope Definition stage.

Interpretation, the last stage, involves the interpretation of results and considers significant environmental issues. It includes an evaluation of the study's completeness, sensitivity and consistency, as well as the formulation of conclusions, limitations and recommendations.

Whole Life Cycle Analysis

Typically, the biggest life cycle carbon impact of buildings is the result of burning fossil fuels to meet the production of building products, transportation and their heating, cooling, hot water and power requirements, otherwise known as their "operational" energy use. As a result, the current emphasis in the sector is on minimising operational energy use through compliance with Part L of the Building Regulations, specifying efficient mechanical and electrical (M&E) systems and using low/zero carbon energy sources ("active" measures). More recently, however, there has been a growing awareness of the fact that buildings are also responsible for manufacturing, transport and construction emissions and, to a lesser extent, maintenance and end of life impacts (together referred to as "embodied" emissions). Indeed, as operational energy efficiencies improve, embodied emissions are becoming an increasingly important component of overall whole life carbon emissions.

The growing awareness of embodied carbon emissions has led to the development of a wide range of industry guidance and tools to support designers in minimising the embodied impacts of their designs (although no regulations, to date). However, it is important that both embodied and operational emissions are considered together in a whole life carbon assessment framework when planning and designing a building or when undertaking a building energy retrofit. This is because there is a trade-off between the two: for example, decreasing operational emissions often requires more material in the envelope, increasing embodied emissions, and vice versa. Therefore, it is not best practice to manage the embodied (or operational) carbon of a building in isolation, since this could result in the inadvertent shifting of environmental burdens to life cycle stages not considered in the analysis. A whole life carbon assessment avoids this problem and ensures that the overall, whole life impact of the building is minimised.

Product Stage

(Modules A1–A3): covering the "cradle-to-gate" processes for the materials and services used during the construction phase of a building project. "Cradle" refers to the raw material resources used, while "gate" refers to the end of the manufacturing process, where the product arrives at the factory "gate" for transport and delivery to the site.

Construction Process Stage

(Modules A4–A5): covering the processes (including transport) from the factory gate of the construction products to the completion of the construction work.

Use Stage

(Modules B1–B7): covers the period from the completion of the construction to the time when the building is being deconstructed/demolished. It includes repair, refurbishment, replacement and operation (heating, cooling, lighting, etc.). B7 deals with operational water use and is not further considered here, since this lies outside the scope of this report.

End of Life Stage

(Modules C1–C4): covers the decommissioning of the building, deconstruction, transport and reuse, recovery and recycling.

Beyond the Life

(Module D): quantifying the net environmental benefits or loads resulting from reuse, recycling and recovery for future use.

Embodied Carbon

The UK Construction Products Association's *A Guide to Understanding the Embodied Impacts of Construction Products* (Construction Products Association, 2012) describes how construction products release the emissions that result in embodied carbon. It summarises how embodied carbon is measured using standard product category rules (based on EN 15804), leading to comparable Environmental Product Declarations (EPDs). An EPD presents the environmental impacts of a product over its life cycle, expressed per functional unit of product (e.g. m³ of concrete or kg of steel). A number of impacts are reported including global warming potential (GWP, expressed in mass of carbon dioxide equivalent – $KgCO_2$eq.), acidification and eutrophication amongst others. It is prepared and presented by a manufacturer according to standard rules (EN 15804) and is independently verified prior to publication.

There are a number of stand-alone guidance documents and tools available which deal exclusively with estimating embodied emissions in buildings. For example, the Low Energy Transformation Initiative Embodied Carbon Primer (LETI, 2020) is aimed at UK building professionals and provides practical guidance on best practice for reducing embodied carbon in UK buildings. The approach involves taking an inventory of all construction materials, obtaining relevant EPD data, with the product of these summing to give the total embodied emissions. The Institution of Structural Engineer's document "How to Calculate Embodied Carbon" provides guidance on estimating the embodied carbon in structures, with a focus on the structural frame and foundations (Gibbons and Orr, 2022). The Royal Institution of Chartered Surveyors *Whole Life Carbon Assessment for the Built Environment* (RICS, 2017) is aimed at building professionals and provides a standard assessment procedure and reporting structure for the implementation of the EN 15978 methodology and is relevant to the UK, although geographic adjustments are highlighted allowing it to be applied in other countries. LETI is aimed at a general audience and describes a whole house retrofit planning methodology using a proforma approach with five different archetypes and various depths of retrofit. It provides practical guidance on moisture and ventilation risks of retrofit and addresses historic buildings.

The IStructE guide provides a common set of embodied carbon calculation principles for the structural engineering community to follow. The assessment framework proposed here follows the EN 15978 standard, is aimed at building professionals and provides WLCA guidance for buildings, with a particular focus on traditional buildings. It includes accounting for material maintenance and replacement over the building's reference service period (60 years) as well as including end of life assumptions.

The above documents summarise methods for estimating embodied emissions. The approaches involve obtaining inventories of building materials from, for example, outline designs, drawings or bills of quantities (depending on the design stage) and combining these with material carbon factors from software, published databases, or EPDs. They describe the need for additional data and assumptions regarding transport distances, maintenance and replacement rates, and end of life treatment to estimate embodied emissions.

Operational Carbon

In addition to estimating embodied emissions, the whole life carbon assessment of building retrofit designs requires the estimation of operational emissions. The literature describes how this first involves estimating the annual energy requirements for a building (e.g. space heating/cooling, hot water and lighting/small power), the quantities of fuels (e.g. gas, oil, electricity) required to meet these requirements. These estimations are combined with a fuel carbon emissions factor and primary energy factors to give annual carbon emissions, which are then scaled up to whole life operational emissions.

ISO 13790:2008 Energy Performance of Buildings – Calculation of energy use for space heating and cooling provides a method to assess the space heating and cooling requirements of buildings (including effects of solar gains) and can be used for comparing the energy performance of various building design alternatives. It provides methods for simple hourly, monthly or dynamic simulation, allowing for building zoning using either quasi-steady-state or dynamic methods. The standard can be used to integrate the impact of heating, ventilation and air conditioning (HVAC) systems on building energy requirements. The Dwelling Energy Assessment Procedure (DEAP) is the Irish domestic dwelling EPBD calculation methodology which is described in the I.S. EN ISO 52000 set of standards (I.S. EN ISO 52000-1:2017, I.S. EN ISO 52003-1:2017, I.S. EN ISO 52010-1:2017, I.S. EN ISO 52016-1:2017, and I.S. EN ISO 52018-1:2017). It is based on I.S. EN 13790 and estimates the energy end uses for space heating, ventilation, water heating and lighting, net of any generation technologies such as renewable energy produced on site. It also estimates carbon dioxide emissions and uses standardised occupancy patterns and comfort settings. Non-domestic Energy Assessment Procedure (NEAP) is the equivalent methodology for non-domestic buildings.

There is no standard operational energy demand assessment method which is consistently used in LCA studies. RICS refers to the use of "SBEM3 (Simplified Building Information Model) and/or dynamic thermal simulation, energy calculations according to CIBSE TM54, etc.", thus leaving it open to the assessor which method to employ. The LETI Climate Emergency Retrofit Guide (LETI, 2021) suggests PHPP4 or, as an alternative, SAP modelling (very similar to Ireland's DEAP). This study uses the DEAP methodology to calculate annual operational energy use which is compliant with the EPBD. An advantage to using DEAP (or NEAP) in Ireland is that these are the methods which are most likely to be implemented as part of any energy retrofitting project and their use is mandatory for building regulations compliance and accessing government and private financial support.

Conclusion

Adaptive Reuse of Existing Buildings

In most if not all countries there are many buildings lying idle, not in use, some derelict and quickly going into a ruinous state. No matter what condition, the adaptive reuse of an existing building can make a big difference to our carbon footprint. The Historic England Study (mentioned earlier) "Understanding Carbon in the Built Environment" showed reusing an existing building can save decades of "Carbon" this includes the "neutral" embodied carbon lost in demolition, the transport of that lost carbon but more importantly the up-front Carbon Intense new construction that tends to replace the old. The construction industry needs to really look at itself and consider the whole life cycle cost of how and what we build and adaptive reuse can largely contribute to the reduction of the industries carbon footprint. Research and studies like the ones mentioned in this chapter are contributing to the science behind reducing our carbon emissions but we now need to implement new practices in adaptive reuse of older buildings, sympathetically energy upgrade our built heritage and introduce new low carbon methods of constructing our new buildings.

We need a mind change from just evaluating "Operational Energy" to evaluating the "Whole Life Cycle Analysis" of the construction industry as a whole. Embodied carbon must be taken into full account when assessing the cost of an adaptive reuse. Many countries and the latest EU Green Initiative are promoting the importance of adaptive reuse and, in many cases, offering generous grants towards achieving this.

Reference List

Adams, K., & Hobbs, G. (2017). *Material resource efficiency in construction: Supporting a circular economy*. BRE Press.

Construction Products Association (2012). *A guide to understanding the embodied impacts of construction products*. CPA.

Department of the Environment, Climate and Communications (2024). Climate Action Plan 2024.

Duffy, Aidan; Nerguti, A; Engel Purcell, C; & Cox, P. (2019). *Understanding carbon in the built environment*. Historic England. https://historicengland.org.uk/content/docs/research/understanding-carbon-in-historic-environment/

Elefante, C. (2007). The greenest building is … one that is already built. *Forum Journal, 21*(4), 26–38.

European Commission (2021). Proposal for a Directive Of The European Parliament and of The Council on the Energy Performance of Buildings (recast) (No. COM(2021) 802 final). Brussels, Belgium.

Gibbons, O. P., & Orr, J. J. (2022). *How to calculate embodied carbon* (Second edition). Institution of Structural Engineers, London.

LETI (2021). *Climate emergency design guide*. LETI, London.

LETI (2020). *LETI embodied carbon primer*. LETI, London.

RICS (2017). *Whole life carbon assessment for the built environment*. RICS, UK.

UK GBC (2019). *Circular economy guidance for construction clients*. UK Green Building Council, London.

3

UNDERSTANDING THE MICRO-CREDENTIALS OF EMBODIED CARBON FROM A MATERIALS SCIENCE PERSPECTIVE

Deepankar Kumar Ashish and Riccardo Maddalena

Introduction

Overview of Embodied Carbon

With a rising climate emergency, decarbonising the built environment has become a critical aspect for mitigating climate change where the construction sector alone is responsible for approximately 37% of global carbon emissions (United Nations Environment Programme, 2023). In line with the Paris Agreement, there is an urgent call to limit global warming to well below 2°C, which emphasises the need to reduce greenhouse gas (GHG) emissions from construction activities (Lu et al., 2024). In response to the changing global climate, many countries are developing strategies to achieve carbon neutrality, the secretary general of the United Nations has declared *"climate emergency in all the countries until carbon neutrality is reached."* The UK pledged to achieve net zero carbon emissions by 2050, whereas China targeted 2060 to reach net zero emissions (Grainger & Smith, 2021). Founded in 1993, World Green Building Council (WorldGBC) has led the global commitment to achieving a net zero goal (*World Green Building Council*, 2023), primarily focusing on operational carbon released from the built environment. However, the role of embodied carbon was overlooked contributing to 11% of global carbon emissions. Often referred to as upfront carbon, these emissions result from the manufacturing, installation, maintenance, and disposal of building materials and infrastructure (Cao, 2017). As operational carbon is gradually addressed through energy-efficient technologies, embodied carbon continues to grow and demands immediate attention (WorldGBC, 2019). Embodied carbon emissions are affected by many factors such as type and volume of structures installed, materials used and their associated carbon intensity of manufacturing processes to the modes

DOI: 10.4324/9781003527404-4

and distances by which materials are transported and the processes by which materials are constructed, maintained, and finally removed and treated at end of life (WorldGBC, 2019).

With growing awareness of building products and futuristic goals, the focus towards reducing embodied carbon has been adopted in many parts of the world (GBCA & thinkstep-anz, 2021). Worldwide research has highlighted two potential strategies that can contribute to reducing embodied carbon:

- Reducing material demand by extending the lifespan of buildings and limiting new construction. This approach emphasises retrofitting and adaptive reuse of existing structures (Lu et al., 2024).
- Developing cleaner materials, which involves initiatives like fuel switching, improving energy efficiency, adopting a circular economy, and advancing carbon capture and storage technologies (Lu et al., 2024).

Conservation of buildings has demonstrated the potential for reducing embodied carbon by extending the lifespan of a structure. Preserving and retrofitting existing structures involves upgrading older buildings for modern use while maintaining their original structure, which minimises the need for new, carbon-intensive materials such as steel, concrete, and glass (Besana & Tirelli, 2022). This is particularly relevant in urban areas where demolition and new construction are prevalent, resulting in significant embodied carbon emissions from both material production and waste generation (Myint & Shafique, 2024). According to a report published in *RMI*, retrofitting and adaptive reuse projects can save between 50% and 75% of embodied carbon compared to constructing a new building (Rosenbloom et al., 2023), in addition it can contribute to obtaining operational efficiency by upgrading insulation, lighting, and water systems. This will optimise the energy performance, further aligning with global sustainability objectives like the United Nations Sustainable Development Goals (SDGs).

In terms of carbon-accounting, the development of low-carbon materials alternative to traditional materials has brought a revolution. Since construction materials like concrete, steel, and aluminium are responsible for a large share of global carbon emissions, focusing on the reduction of embodied carbon is crucial for sustainable development. Concrete, which accounts for 8% of global carbon emissions, is an area of intense research (Maddalena et al., 2018). The greenhouse gas emissions associated with building materials are projected to increase at a rate of 0.7% per year from 2020 to 2060, growing from 3.5 gigatonnes to 4.6 gigatonnes of CO_2 equivalent (CO_{2e}) per year over that period. This growth reflects the continued impact of the construction industry on global emissions, highlighting the importance of reducing embodied carbon in building materials (Zhong et al., 2021). Innovations such as geopolymer concrete, which uses industrial waste by-products like fly ash or steel slag instead of energy-intensive

Portland cement, have shown promise in reducing emissions by up to 80% (Almutairi et al., 2021). Similarly, engineered wood products like cross-laminated timber (CLT) offer carbon sequestration benefits while providing comparable strength to steel and concrete (Younis & Dodoo, 2022). By focusing on both traditional high-carbon materials and innovative low-carbon alternatives, this chapter will provide a comprehensive view of how material science can be harnessed to reduce the environmental impact of the built environment.

The Role of Materials in Embodied Carbon

Materials as a Key Contributor

Estimates suggest that 40–50% of resources extracted across the world are used for infrastructure, construction, and housing (Brady & Kawamura, 2021). Within this context, concrete stands out as a material with a significant carbon footprint. While cement comprises only about 10% of concrete by volume, it accounts for approximately 75–90% of its total embodied carbon (ARUP, 2023). This disproportionate impact is primarily due to the energy-intensive process of cement production, where raw materials are heated to around 1450°C to form Portland cement clinker. Of the CO_2e emissions released during this process, 40% come from fuel combustion, while 60% arise from the chemical reactions involved. For each tonne of cement produced, approximately 860 kg of CO_2e is emitted into the atmosphere (Maddalena et al., 2018). Other than cement another major component is aggregate. Aggregates are naturally occurring materials and only account for 1–5% emissions, but their extremely low formation relative to its huge demand threatens their long-term availability (UNEP, 2022). In the pursuit of low-carbon infrastructure, understanding the role of aggregates in embodied carbon of concrete is essential. A study by Flower and Sanjayan highlights that the type of aggregate significantly impacts CO_{2e} emissions. For instance, granite/hornfels production emits 0.0459 t CO_{2e}/tonne, while basalt and concrete-sand contribute 0.0357 t CO_{2e}/tonne and 0.0139 t CO_{2e}/tonne, respectively (Flower & Sanjayan, 2007). Additionally, Purnell and Black explored different aggregate types and cement strength classes to understand their effect on the carbon footprint. The study suggested that shifting from uncrushed to crushed aggregates can reduce CO_{2e} by 9%, and using higher cement strength classes further lowers CO_{2e} by 7%. Although the reductions are moderate, such optimisations in aggregate selection contribute meaningfully to reducing the overall embodied carbon in concrete structures (Purnell & Black, 2012). The effect of large size aggregates was evaluated in a study conducted by Gan et al., where eco-cement was used produced from municipal solid waste and the results showed a 17% reduction in embodied carbon (J. Gan et al., 2015). Replacement of natural aggregate with recycled aggregate in proportions of 50% and 100% has been shown to increase carbon emissions, however, the concrete mixes with 100% recycled aggregates and 25% fly ash showed lower carbon emissions relative to natural aggregate (Sabău et al., 2021).

In some cases, admixtures are added to concrete to increase its workability and reduce the amount of water required in the mix, increasing the overall strength of the concrete by reducing permeability without increasing the cement ratio which is carbon intensive. Using a plasticising admixture is generally accommodated to consume less water for required consistence where the cement content of the concrete can be reduced to maintain an equal water/cement ratio. The concrete will therefore have lower embodied carbon but with equal strength and durability. The admixtures typically have a small contribution to carbon emission lying between 0.5–3.0 kg CO_{2e}/kg (ARUP, 2023).

Alongside cement, steel also significantly contributes to carbon emissions due to its widespread use and intensive manufacturing. Reinforced concrete, which combines cement, aggregate, and steel rebars, is used at a massive annual rate exceeding 10 billion tonnes. This makes concrete buildings responsible for 90% of their total embodied carbon. Several studies have explored the embodied carbon of both plain and reinforced concrete (V. J. L. Gan et al., 2019). In a study conducted by Zhang et al. (2014), embodied carbon value for plain concrete was analysed as 425 $kgCO_{2e}$/m³ (Zhang et al., 2014). Additionally, Flower and Sanjayan (2007) investigated the relationship between concrete's compressive strength and its embodied carbon, concluding that stronger concrete (with compressive strength between 25–32 MPa) has a slightly higher carbon footprint, in the range of 225–322 $kgCO_{2e}$/m³ (Flower & Sanjayan, 2007).

Other building materials like aluminium, glass, and insulation further increase the embodied carbon. While windows and insulation are typically associated with reducing operational energy consumption by improving thermal performance, their embodied energy is also substantial. The choice of materials in window systems affects the embodied carbon. For instance, double-glazed aluminium windows with no recycled content have a much higher embodied energy (2,219 kWh) compared to single-glazed wooden windows (84.1 kWh). The number of panes, the type of glass, and the frame material all contribute to both the embodied and operational energy profiles of buildings. Considering the overall life cycle of building components, minimising embodied carbon in windows and insulation materials becomes essential for an integrated approach to energy efficiency (Azari & Abbasabadi, 2018).

Materials Science Approach to Reducing Embodied Carbon

Key Factors Influencing Embodied Carbon in Concrete

Although embodied carbon in concrete can vary significantly, the primary sources of carbon emissions are ordinary Portland cement (OPC) and steel reinforcement (V. J. L. Gan et al., 2019). The carbon emissions of concrete are primarily influenced by its cement content; as concrete strength increases, the amount of cement needed to maintain workability and compaction also rises, leading to a higher CO_2 emission. For a required strength, the carbon emissions can vary up to threefold depending on cement content (Purnell & Black, 2012).

While increasing cement content raises emissions, the relationship between cement content and strength is further complicated by the water-to-binder ratios. Thus, increasing concrete strength does not always result in a proportional rise in cement content or CO_2 emissions. Studies suggest that optimal structural performance with minimal embodied carbon is achieved at around 60 MPa. Below this strength, less cement is required, but more concrete volume is needed to meet structural demands. On the other hand, stronger concretes reduce material use but demand more cement, thereby increasing carbon emission. However, utilising concrete at its optimal strength can reduce CO_2 emissions by 40%. The workability also has a significant effect on embodied carbon, the emission can be seen reducing with the slump class from 60–180 mm to 0–10 mm by 35±1%. Additionally, incorporating superplasticisers which enhance workability without increasing water content allows for a reduction in the water-to-binder ratio. This adjustment further decreases binder content, leading to a 26±1% reduction in embodied CO_{2e}. These changes provide substantial carbon savings with minimal compromise on the mix's performance (Mergos, 2024).

Role of Supplementary Cementitious Materials (SCMs)

Supplementary cementitious materials (SCMs) can reduce embodied carbon by partially replacing OPC, some of the SCMs that are accommodated in replacement to OPC are ground granulated blast furnace slag (GGBS), limestone, pulverised fuel ash (PFA) and silica fume. For example, a standard C25/30 concrete mix in the UK has an embodied carbon of 0.12 $kgCO_{2e}$/kg when using only OPC. However, with a 70% replacement of cement by GGBS, the embodied carbon drops to 0.06 $kgCO_{2e}$/kg, showing a 50% reduction (Mergos, 2024). SCMs contribute to lower emissions not only during production but also improve the durability and performance of concrete, thus further reducing the need for material over time.

Many past studies replaced OPC with SCMs such as fly ash or GGBS and suggested a decrease in carbon emissions by 20–30% (García-Segura et al., 2014). In another study, plain concrete with 100% OPC was compared with other comparable mix-designs having geo-polymers, showing that samples containing geo-polymers produced 8% less emissions that is 320 $kgCO_{2e}$/m³ relative to OPC concrete with 354 $kgCO_{2e}$/m³ (Habert et al., 2011; Turner & Collins, 2013). The replacement of 5% cement with silica fume reduces total carbon emission to 390.232 $kgCO_{2e}$/m³, a reduction of 4.75% as compared to the control mix which produced 409.68 $kgCO_{2e}$/m³. With an increment increase in silica fume the carbon emissions were observed decreasing, however, replacement of 15% silica fume and 20% of fly ash further reduced the CO_2 emission of 273.784 $kgCO_{2e}/m^3$, a reduction of 33.17% as compared to 409.68 $kgCO_{2e}/m^3$. The mix was observed to achieve optimal strength properties owing to the pore filling effect of both materials (Lashari et al., 2023). Purnell and Black evaluated that replacing 40% OPC can lower carbon emissions by approximately 35±1%, moreover, since PFA is 30% less dense than cement, replacing 40% of cement with PFA requires an increase of about 13% in the total binder mass to maintain the same paste fraction, ultimately resulting in a 35% reduction in cement content (Purnell & Black, 2012).

Steel Reinforcement and Embodied Carbon

The carbon footprint of steel reinforcement (rebars) is primarily linked to the energy required to process steel scrap and iron ore. On average, 1 kg of steel reinforcement generates approximately 2 $kgCO_{2e}$, but this can vary between 0.4 $kgCO_{2e}$ and 4 $kgCO_{2e}$ depending on factors such as scrap content and steel production route. The basic oxygen furnace (BOF), powered by fossil fuels, allows for only up to 30% scrap content, while the electric arc furnace (EAF), powered by electricity, can incorporate up to 100% recycled steel. Rebars produced through the EAF method with higher scrap content emit significantly lower embodied carbon relative to BOF-produced steel (Mergos, 2024). Similarly, Harrison et al. (2010) offered an estimate of 0.13 $kgCO_{2e}$ per kilogram for plain concrete, increasing to 0.24 $kgCO_{2e}$ per kilogram when steel reinforcement is added, which raised emissions by approximately 2% (Harrison et al., 2010). Miller et al. (2015) expanded on these findings, showing that the embodied carbon of reinforced concrete ranges from 150 to 520 $kgCO_{2e}$ per cubic meter, depending on factors like the mix proportions, the geometry of the building element, and its design age (Miller et al., 2015). On comparison of concrete and steel structural systems for a typical bay of 7.5 m by 7.5 m, embodied energy was found to be 42GJ for concrete structures and 55GJ for steel structures. Additionally, 10–15% of the total embodied energy is consumed in the foundations (Cole & Kernan, 1996). It is necessary to quantify the relationship between material production and embodied carbon, providing clearer insights for material manufacturers and designers (Gursel et al., 2014).

Integrated Strategies for Sustainable Building

Increasing concrete compressive strength by 100% can reduce material usage by 36% and cut embodied carbon by 11%. Moreover, using SCMs like fly ash and GGBS can further lower emissions. Using SCMs like fly ash (35%) can cut embodied carbon by 16%, while a 75% GGBS replacement can result in a 31% reduction. Additionally, using 100% recycled steel scrap for reinforcement can decrease total building embodied carbon by 39%. Other strategies, like employing eco-cement, can reduce embodied carbon by 14%, and using 40 mm aggregates can lower it by 9%, underscoring the need for an integrated approach to sustainable building practices (Gursel et al., 2014). In recent years, researchers worldwide are increasingly finding new alternatives for reducing embodied carbon that include wood, bamboo, and other renewable materials with steel or concrete composites. In a study conducted by Becker et al., a steel-timber system was compared to a steel-concrete system and revealed that the GWP (global warming potential) of a steel-timber system without topping slab was 83 $kgCO_{2e}/m^2$ relative to steel-concrete where GWP was 127 $kgCO_{2e}/m^2$, approximately 35% lower than a steel-concrete combination (Becker et al., 2023). Liu developed a concrete-wood structure that was able to give satisfactory seismic performance and reduced carbon emission during production and construction stage (Liu, 2020). Zhao et al. compared a steel timber composite frame structure (STF) scheme and section steel frame structure (SSF) scheme and observed that STF had a 10.22% lower carbon emission relative to SSF. Souaid et al., suggested that using timber as a main material can reduce embodied carbon by 22–24% relative to a building using conventional materials (Souaid et al., 2024).

Material Efficiency in Construction

Material efficiency in construction has become essential for reducing both environmental impact and embodied carbon, prompting the industry to adopt innovative strategies to minimise waste, optimise resources, and advance sustainable practices. Modular and prefabricated construction techniques, for example, offer efficient material usage, producing minimal waste (Rosenbloom et al., 2023). Studies indicate that building information modelling (BIM) further enhances structural modelling by creating highly detailed 3D models. It allows for precise measurement and material estimation, reducing excess use and minimising construction waste (Azhar, 2011). BIMs precise material estimation can reduce construction waste by up to 20%, while reduction in carbon emissions can be reduced by up to 50%.

In modular and prefabricated construction, self-contained building units are manufactured off-site, including various structural components such as precast concrete columns, panelised wall systems with exterior/interior walls, floors, and roofs. These incorporate mechanical, electrical, and plumbing systems where the most extensive system is called modular construction. According to Greer and Horvath, modular housing can achieve greenhouse gas reductions of 4–20% relative to conventional on-site housing units (Greer & Horvath, 2023). Teng et al., evaluated 27 case studies of prefabricated systems and revealed a wide range of embodied carbon emissions ranging from 105–864 $kgCO_{2e}/m^2$, on average case studies showed a 15.6% reduction in embodied energy relative to on-site construction projects (Teng et al., 2018).

Additionally, 3D printing is an emerging technology in construction, building objects layer by layer, allowing precise design. Unlike traditional manufacturing methods, 3D printing creates three-dimensional objects based on a digital model. The average contribution of 3D-printed houses towards greenhouse gas emissions is 58 $kgCO_{2e}/m^2$ for the 3D-printed houses, and 147 $kgCO_{2e}/m^2$ for conventional houses (Rossi et al., 2024).

Conclusions

Addressing embodied carbon that accounts for considerable portion of a building's total carbon footprint is essential for achieving global climate targets. Through the analysis of materials like concrete, steel, and innovative low-carbon alternatives, it becomes evident that material science holds the potential to mitigate carbon emissions. In this chapter, the concept of embodied carbon is thoroughly examined, understanding the intricate relationship between material selection and carbon footprint. Key findings indicate that while cement is a primary driver of carbon emissions, strategic modifications in material choices such as incorporation of SCMs and recycled aggregates can contribute to lowering carbon emissions significantly. The studies cited demonstrate that effective material management and optimised design not only enhance structural performance but also contribute to substantial carbon reductions.

Moreover, the integration of advanced technologies like BIM and adoption of modular construction techniques offer promising avenues for minimising waste and improving efficiency. By emphasising the use of renewable materials and innovative construction methods, stakeholders can reduce the carbon intensity of buildings and promote a more sustainable future. This chapter underscores the need for a collaborative approach among architects, engineers, and policy makers to prioritise low-carbon solutions in construction. As industry moves towards a sustainable paradigm, the insights presented here serve as a foundational guide for selecting materials that align with global sustainability objectives, paving the way for a greener built environment.

Future Directions to Embodied Carbon Reduction

Although strategies like building design and construction planning have been studied for reducing embodied carbon, there remains a lack of systematic analysis on how variations in reinforced concrete mix designs at the production stage affect total embodied carbon in buildings. Development and adoption of carbon-neutral materials, such as bio-based composites and innovative concrete alternatives can make sustainable materials viable and accessible to industry. Modular and prefabricated techniques offer significant potential to minimise embodied carbon through reduced waste, efficient material use, and shortened build times. Research and development should focus on expanding these methods and creating standardised, low-carbon modular solutions for various building types.

Reference List

Almutairi, A. L., Tayeh, B. A., Adesina, A., Isleem, H. F., & Zeyad, A. M. (2021). Potential applications of geopolymer concrete in construction: A review. *Case Studies in Construction Materials, 15*, e00733. https://doi.org/10.1016/J.CSCM.2021.E00733

ARUP (2023). *Embodied Carbon: Concrete.*

Azari, R., & Abbasabadi, N. (2018). Embodied energy of buildings: A review of data, methods, challenges, and research trends. *Energy and Buildings, 168*, 225–235. https://doi.org/10.1016/J.ENBUILD.2018.03.003

Azhar, S. (2011). Building information modeling (BIM): Trends, benefits, risks, and challenges for the AEC industry. *Leadership and Management in Engineering, 11*(3), 241–252. https://doi.org/10.1061/(ASCE)LM.1943-5630.0000127/ASSET/D25A4476-6404-41B6-88CA-7C40041AA486/ASSETS/IMAGES/LARGE/5.JPG

Becker, I., Anderson, F., & Phillips, A. R. (2023). Structural design of hybrid steel-timber buildings for lower production stage embodied carbon emissions. *Journal of Building Engineering, 76*, 107053. https://doi.org/10.1016/J.JOBE.2023.107053

Besana, D., & Tirelli, D. (2022). Reuse and retrofitting strategies for a net zero carbon building in Milan: An analytic evaluation. *Sustainability 2022, 14*(23), 16115. https://doi.org/10.3390/SU142316115

Brady, C., & Kawamura, S. (2021). *An integrated approach to a sustainable built environment: The co-benefits of resources & circularity.* WorldGBC. https://worldgbc.org/article/an-integrated-approach-to-a-sustainable-built-environment-the-co-benefits-of-resources-circularity/. Accessed 10.11.24.

British Standards (2006). BS EN ISO 14044:2006 – Environmental management — Life cycle assessment — Requirements and guidelines. In *British Standards.* British Standards.

British Standards Institution (2006). BS EN ISO 14040:2006 – Environmental management — Life cycle assessment — Principles and framework. In *British Standards* (Issue August 2006).

Cao, C. (2017). Sustainability and life assessment of high strength natural fibre composites in construction. *Advanced High Strength Natural Fibre Composites in Construction,* 529–544. https://doi.org/10.1016/B978-0-08-100411-1.00021-2

Cole, R. J., & Kernan, P. C. (1996). Life-cycle energy use in office buildings. *Building and Environment, 31*(4), 307–317. https://doi.org/10.1016/0360-1323(96)00017-0

Flower, D. J. M., & Sanjayan, J. G. (2007). Green house gas emissions due to concrete manufacture. *International Journal of Life Cycle Assessment, 12*(5), 282–288. https://doi.org/10.1065/LCA2007.05.327/METRICS

Gan, J., Cheng, J. C. P., & Lo, I. M. C. (2015). A systematic approach for low carbon concrete mix design and production. *3rd International Conference on Civil Engineering, Architecture and Sustainable Infrastructure.* https://hdl.handle.net/1783.1/77688. Accessed 10.11.24.

Gan, V. J. L., Cheng, J. C. P., & Lo, I. M. C. (2019). A comprehensive approach to mitigation of embodied carbon in reinforced concrete buildings. *Journal of Cleaner Production, 229*, 582–597. https://doi.org/10.1016/J.JCLEPRO.2019.05.035

García-Segura, T., Yepes, V., & Alcalá, J. (2014). Life cycle greenhouse gas emissions of blended cement concrete including carbonation and durability. *International Journal of Life Cycle Assessment, 19*(1), 3–12. https://doi.org/10.1007/S11367-013-0614-0/TABLES/8

GBCA, & thinkstep-anz. (2021). *Embodied carbon & embodied energy in Australia's buildings.*

Grainger, A., & Smith, G. (2021). The role of low carbon and high carbon materials in carbon neutrality science and carbon economics. *Current Opinion in Environmental Sustainability*, *49*, 164–189. https://doi.org/10.1016/J.COSUST.2021.06.006

Greer, F., & Horvath, A. (2023). Modular construction's capacity to reduce embodied carbon emissions in California's housing sector. *Building and Environment*, *240*, 110432. https://doi.org/10.1016/J.BUILDENV.2023.110432

Gursel, A. P., Masanet, E., Horvath, A., & Stadel, A. (2014). Life-cycle inventory analysis of concrete production: A critical review. *Cement and Concrete Composites*, *51*, 38–48. https://doi.org/10.1016/j.cemconcomp.2014.03.005

Habert, G., D'Espinose De Lacaillerie, J. B., & Roussel, N. (2011). An environmental evaluation of geopolymer based concrete production: Reviewing current research trends. *Journal of Cleaner Production*, *19*(11), 1229–1238. https://doi.org/10.1016/J.JCLEPRO.2011.03.012

Harrison, G. P., Maclean, E. (Ned) J., Karamanlis, S., & Ochoa, L. F. (2010). Life cycle assessment of the transmission network in Great Britain. *Energy Policy*, *38*(7), 3622–3631. https://doi.org/10.1016/J.ENPOL.2010.02.039

Lashari, A. R., Kumar, A., Kumar, R., & Rizvi, S. H. (2023). Combined effect of silica fume and fly ash as cementitious material on strength characteristics, embodied carbon, and cost of autoclave aerated concrete. *Environmental Science and Pollution Research*, *30*(10), 27875–27883. https://doi.org/10.1007/S11356-022-24217-9/FIGURES/6

Liu, J. (2020). *Mechanical behavior and carbon emission analysis of a concrete mega structure filled with timber structure* [Masters' degree]. Central South University of Forestry and Technology.

Lu, H., You, K., Feng, W., Zhou, N., Fridley, D., Price, L., & de la Rue du Can, S. (2024). Reducing China's building material embodied emissions: Opportunities and challenges to achieve carbon neutrality in building materials. *IScience*, *27*(3), 109028. https://doi.org/10.1016/J.ISCI.2024.109028

Maddalena, R., Roberts, J. J., & Hamilton, A. (2018). Can Portland cement be replaced by low-carbon alternative materials? A study on thermal properties and carbon emissions of innovative cements. *Journal of Cleaner Production*, *186*, 933–942. https://doi.org/10.1016/j.jclepro.2018.02.138

Mergos, P. E. (2024). Structural design of reinforced concrete frames for minimum amount of concrete or embodied carbon. *Energy and Buildings*, *318*, 114505. https://doi.org/10.1016/J.ENBUILD.2024.114505

Miller, S. A., Horvath, A., Monteiro, P. J. M., & Ostertag, C. P. (2015). Greenhouse gas emissions from concrete can be reduced by using mix proportions, geometric aspects, and age as design factors. *Environmental Research Letters*, *10*(11), 114017. https://doi.org/10.1088/1748-9326/10/11/114017

Myint, N. N., & Shafique, M. (2024). Embodied carbon emissions of buildings: Taking a step towards net zero buildings. *Case Studies in Construction Materials*, *20*, e03024. https://doi.org/10.1016/J.CSCM.2024.E03024

Purnell, P., & Black, L. (2012). Embodied carbon dioxide in concrete: Variation with common mix design parameters. *Cement and Concrete Research*, *42*(6), 874–877. https://doi.org/10.1016/J.CEMCONRES.2012.02.005

Rosenbloom, E., Magwood, C., Clark, H., & Olgyay, V. (2023). *Transforming existing buildings from climate liabilities to climate assets.* https://rmi.org/insight/transforming-existing-buildings-from-climate-liabilities-to-climate-assets/. Accessed 10.11.24.

Rossi, C., Reitemeyer, F., Heidrich, O., & Rybski, D. (2024). Comparison of embodied carbon of 3D-printed vs. conventionally built houses. *Findings*. https://doi.org/10.32866/001c.89707

Sabău, M., Bompa, D. V., & Silva, L. F. O. (2021). Comparative carbon emission assessments of recycled and natural aggregate concrete: Environmental influence of cement content. *Geoscience Frontiers, 12*(6), 101235. https://doi.org/10.1016/J.GSF.2021.101235

Souaid, C., ten Caat, P. N., Meijer, A., & Visscher, H. (2024). Downsizing and the use of timber as embodied carbon reduction strategies for new-build housing: A partial life cycle assessment. *Building and Environment, 253*, 111285. https://doi.org/10.1016/J.BUILDENV.2024.111285

Teng, Y., Li, K., Pan, W., & Ng, T. (2018). Reducing building life cycle carbon emissions through prefabrication: Evidence from and gaps in empirical studies. *Building and Environment, 132*, 125–136. https://doi.org/10.1016/J.BUILDENV.2018.01.026

Turner, L. K., & Collins, F. G. (2013). Carbon dioxide equivalent (CO_2-e) emissions: A comparison between geopolymer and OPC cement concrete. *Construction and Building Materials, 43*, 125–130. https://doi.org/10.1016/J.CONBUILDMAT.2013.01.023. Accessed 10.11.24.

UNEP (2022). *Sand and sustainability: 10 strategic recommendations to avert a crisis.* https://unepgrid.ch/en/resource/2022SAND. Accessed 10.11.24.

United Nations Environment Programme, & Y. C. for E. + A. (2023). *Building materials and the climate: Constructing a new future.*

World Green Building Council (2023). https://worldgbc.org/. Accessed 10.11.24.

WorldGBC. (2019). *Bringing embodied carbon upfront.*

Younis, A., & Dodoo, A. (2022). Cross-laminated timber for building construction: A life-cycle-assessment overview. *Journal of Building Engineering, 52*, 104482. https://doi.org/10.1016/J.JOBE.2022.104482

Zhang, J., Cheng, J. C. P., & Lo, I. M. C. (2014). Life cycle carbon footprint measurement of Portland cement and ready mix concrete for a city with local scarcity of resources like Hong Kong. *International Journal of Life Cycle Assessment, 19*(4), 745–757. https://doi.org/10.1007/S11367-013-0689-7/TABLES/9

Zhong, X., Hu, M., Deetman, S., Steubing, B., Lin, H. X., Hernandez, G. A., Harpprecht, C., Zhang, C., Tukker, A., & Behrens, P. (2021). Global greenhouse gas emissions from residential and commercial building materials and mitigation strategies to 2060. *Nature Communications 2021, 12*(1), 1–10. https://doi.org/10.1038/s41467-021-26212-z

Part 2

Extrapolations

4

RESTORATION vs REDEVELOPMENT

Embodied Carbon in Conserving a 20th-Century Heritage in India

Jigna Desai and Rajan Rawal

With inputs from Nigar Shaikh for material characterisation studies and Sneha Asrani for embodied energy calculations.

Introduction

The post-independence decades of the 1950s, 60s and 70s in India, and in South Asia, are marked by rigorous building design and construction activities. The economic impetus of the countries was closely linked to urbanisation, industrialisation and establishing institutions that would shape the future. The first generation of architects to get major commissions in India were formally trained outside the country owing to the fact that there was only one institution for architectural education at the time of independence, namely the Sir J. J. School of Architecture, with around 300 architects (Chatterjee, 1985).[1] These decades were ripe with the ideological debates on the architecture that would truly represent Indian identity. The practices of some architects, such as Habib Rehman and Achyut Kanvinde aligned with Modernist ideologies, which came to be known as the 'International Style', while the others were known as 'traditionalists'. In the early 1950s the first Prime Minister of the country Pandit Jawaharlal Nehru, invited Le Corbusier, a French architect, designer, urban planner and one of the pioneers of modernist architecture, to design the city of Chandigarh, giving an indirect political sanction to the new direction in what arguably became the 'Indian Modern'. The 'International Style' of Architecture and Urbanism dominated the production of public architecture in the decades that followed. As a corollary to that, exposed RCC and exposed brick walls became the idiom of the time, and cement, steel and bricks the most used materials as a large-scale upscaling of production followed.

[1] A Department of Architecture was set up at the Delhi Polytechnic of Delhi University in the early 1940s, which was re-named the School of Planning and Architecture (SPA) post-independence. Architecture department at the Indian Institute of Technology (IIT), Kharagpur and at the Maharaja Sayajirao University (MSU), Vadodara were set up in the 1950s.

DOI: 10.4324/9781003527404-6

[2] The dormitories of Indian Institute of Management (IIM) Ahmedabad, built by Louis Kahn, and Sanskar Kendra at Ahmedabad, built by Le Corbusier have been a part of the debate of restoration vs redevelopment for almost a decade.

[3] Please refer to the various debates around conservation of modernist buildings in India as archived in https://architexturez.net, last accessed on 1st November 2024.

[4] The iconic image of the 1950s was Prime Minister Jawaharlal Nehru giving a speech while standing in front of the Bhakra Nangal Dam, an engineering feat that was to bring water to multiple states, and was only possible due to the mouldability of reinforced cement concrete.

The 20th-Century Heritage of India

A number of public institutions built during the decades mentioned above have come to be regarded as a part of a significant movement of post-independence architecture of the region that allowed architects to use the lens of Modernism to liberate themselves from the colonial idioms and simultaneously search for roots (Curtis, 1987). These structures are now more than 60 years old, and as discussed later in the chapter have possibly outlived their material life cycle. Many of the structures are now in a dilapidated condition and are being considered for demolition.[2] Politics of heritage notwithstanding, the reasons usually slated for their demolition are their structural instability and inability to adapt to the requirements of the constantly changing nature of their occupation. The general understanding is that it would be cheaper to redevelop rather than restore.[3] As identified by Croft and Macdonald (2018 pp. 11–13), the global experience of conserving concrete structures suggests that concrete is 'one of the most ubiquitous materials of the twentieth century' and conserving historic concrete which requires sensitive, reversible repair to aim at maximum retention of the historic fabric is particularly challenging. In the Indian context, this is a peculiarly amplified problem considering that the structures built in the two decades after independence were potentially compromised owing to the gap between the demand and supply of building construction material, especially cement. Bapat et al. (2007) note that the price and distribution of cement continued to remain in a controlled regime until 1977, after which the Government of India gradually introduced decontrol mechanisms.

A large demand for public institutions, coupled with the rationing of building materials, led to design experimentations by architects and engineers that became the ethos of architectural production of the time. Architects along with engineers of the day produced large spanned iconic public structures made of exposed brick and concrete, materials ideologically associated with the Modernist movement and the industrial future that India was heading towards.[4] Ahmedabad, a city in western India, owing to the patronage of wealthy and visionary industrialists, became a hub for experiments of 'Indian Modern'. In the decades of the 1940s and 50s, the industrialists who were invested in the growth of the city and the country invited modern master architects, such as Frank Lloyd Wright, Buckminster Fuller, Le Corbusier and Achyut Kanvinde to build their homes and their organisations. In 1960, the state of Gujarat officially was separated from Bombay State and while Gandhinagar was being built as its new capital, the city of Ahmedabad became the place where new educational, social and cultural institutions were being planned. The National Institution of Design (NID) designed by Gira Sarabhai and Gautam Sarabhai, Center for Environmental Planning and Technology (now CEPT University, mentioned as CEPT University in this chapter) designed by B. V. Doshi, and the Indian Institute of Management (IIM) designed by Louis Kahn along with Anant Raje were the most prominent educational institutes built at the time. Each of these institutions built in exposed brick wall and exposed RCC are now eroding and are in dire need of conservation. This is also true for all institutions built at the time, including the Gandhi Ashram Memorial and Sardar Patel Stadium by Charles Correa, Physical Research Laboratory by Achyut Kanvinde, the Lalbhai Dalpatbhai Museum by B. V. Doshi and more. The School of Architecture at CEPT University, the part that was built between 1964 and 1967 is arguably the only building that has been repaired and conserved keeping the values of modernist architecture in mind while upgrading the structure for current and future needs.[5] The chapter takes this building as the primary case study for the analysis of embodied carbon conserved by taking the decision to repair and restore the building.

The School of Architecture at CEPT, a Background

The School of Architecture at Ahmedabad was established in 1962 under the Ahmedabad Education Society that was actively supported by the industrialists of the city. The curriculum and the pedagogy of the school was the idea of openness towards learning from new ideas as well as the traditions of the country, of 'education without doors', articulated by the architect B. V. Doshi – see Figure 4.1.

Design of the school also reflected that in form of free-flowing spaces and movement through them. The school building, along with the CEPT University campus, is now considered to be one of the significant buildings that represent the post-independence Modern movement in India.[6] Construction of the School of Architecture building at the CEPT University started in 1964. It was built on a 22-acre (approximately 9 hectare) site that was formerly a brick kiln. The load bearing structure of the building with 600 mm (six-and-a-half bricks) thick exposed brick walls at 11.10 meters supporting ribbed reinforced cement concrete

[5] The authors were actively involved in the process of conservation of the School of Architecture at the CEPT University. Jigna Desai was appointed as the conservation architect and Rajan Rawal was responsible for designing the upgraded comfort systems of the building.

[6] In the years between 2015 and 2019, international organizations such as DOCOMOMO dedicated to documentation and conservation of buildings, sites and neighbourhoods of the Modern Movement, and International Council for Monuments and Sites (ICOMOS) recognised the campus, especially the School of Architecture as significant 20th-century heritage for the world. This sentiment was also seen in many letters and communications from national cultural organisations as well as prominent architects of the world. The letters are archived at https://archi texturez.net, last accessed on 1st November 2024.

Figure 4.1 The School of Architecture, Ahmedabad.
(Dinesh Mehta photography – 2022)

(RCC) with a unique staggered section allows large teaching spaces (studios) to interact with another while being flexible – see Figure 4.2. The north and the south façades open up for fenestrations and are designed to address the climatic conditions of the place. The south façade has deep vertical fins with pivoted doors that provide adequate shade in summer and allow in the winter sun, the north façade only allows diffused daylight from the elevated opening to create appropriate luminous working conditions for the function to be carried out. Parts of the north and south façades were closed with brick walls having less thickness, which in some cases had a vertical cavity between two walls. As per the earthquake resistance standards of the time, these walls were reinforced horizontally with steel bars at every three feet in height.

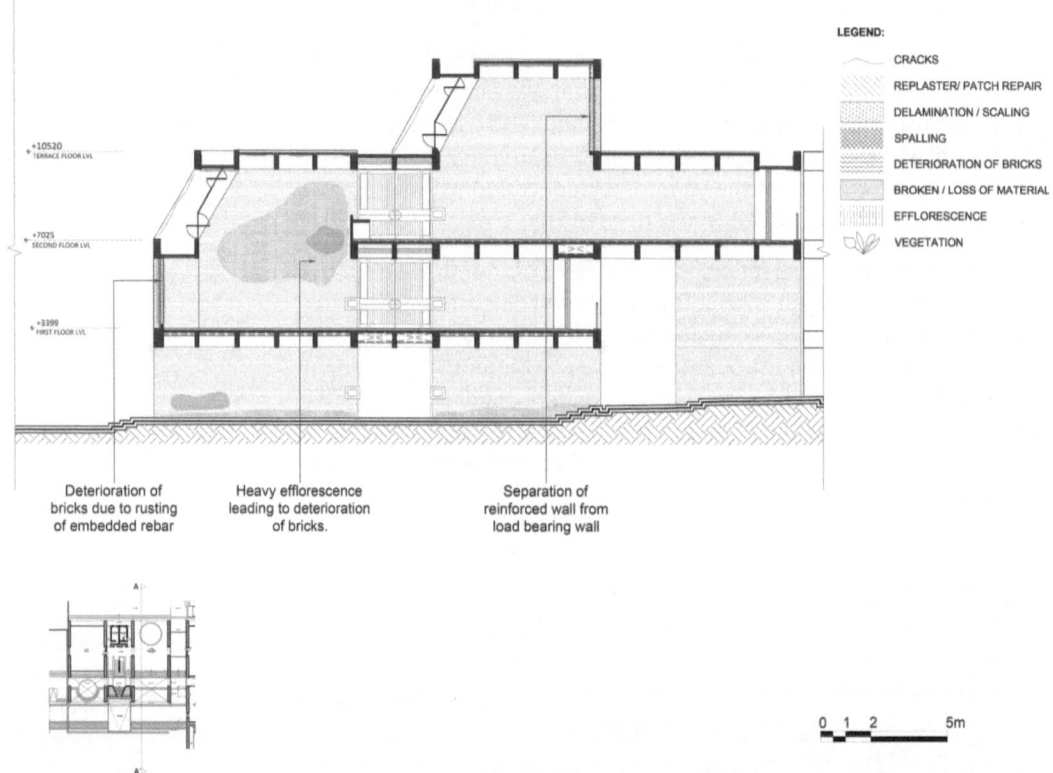

Figure 4.2 Cross section of the School of Architecture Building with preliminary condition assessment. (Sarvesh Alshi 2020)

⁷ This study was carried out by Prof. R. J. Shah a noted structural engineer and a professor at the University. The unpublished report is available at the CEPT archives.

The building has undergone many minor spatial changes and construction repairs, including after the severe earthquake of 2001, none of this, however, altered the structure and the fenestration of the building. Over the years, the condition of the structure, especially the concrete elements, had started to deteriorate. This was noted by the

University management and a systematic study to understand the condition was carried out between 2017 and 2018.[7] These visual observations were then supported with Ultrasonic Pulse Velocity tests, rebound hammer tests (on concrete), density and water absorption tests on the concrete core, crushing strength and porosity tests on bricks, and yield strength on steel members. These partially invasive structural stability tests supported the findings that the load bearing brick walls were mostly in good condition, except for spalling of pointing, efflorescence, and deterioration of surface bricks in some areas. Moisture patches were also found in the load bearing walls pointing towards the cause being the degraded waterproofing, plumbing, and groundwater treatment of the surrounding area. Other than this, diagonal structural cracks were found at two places which needed repairs. It was the RCC elements that had decayed significantly, most of it due to the advance of a carbonation front towards the embedded steel that causes corrosion and triggers cracking and spalling, a condition globally identified as the primary cause of concrete decay (Croft and Macdonald, 2018). In case of the School of Architecture building, the condition of advanced carbonation and corrosion was hastened by leakage of water during monsoons. Structural concrete and fins were either cracked, surface concrete was delaminated, or in some cases had completely deteriorated (see Figure 4.3). Reinforcing steel, also used in the partition walls, had undergone corrosion and had resulted in deformation of these walls. After this extensive study, it was found that less than 120 cubic meters out of the total 655 cubic meters of brickwork (less than 18%) and less than 85 cubic meters out of the total 520 cubic meters of RCC work (less than 16%) was to be removed and restored or repaired.

Figure 4.3 Damaged terrace gutter slab. (CHC 2020)

[8] The team included (in alphabetical order): Apurv Patel, Ashish Jani, Audery Alvarez, Bhargav Tewar, Chirag Dave, Dilip Patel, Jigna Desai, Kalgi Patel, Keyur Patel, Khushi Shah, Lalji Patel, Mehul Shah, Mihir Patel, Nigar Shaikh, R. J. Shah, Rajan Rawal, Sarvesh Alshi, Shweta Maiyatra, Sudeep Vishwakarma, Urnit Kaur.

Strategies for Conserving Brick Walls and Concrete

Conservation of the School of Architecture was taken up between March 2020 and June 2021. The process involved a large team of experts, architects, designers and engineers.[8] The process included repairs and recasting of the RCC structural and non-structural elements, removing and reconstructing brick partition walls, repairing the one major crack in the structural brick wall, replacing the originally operable single glass fenestration with high-performance double-glazed units and fixed glass fenestration towards the north side and minor repairs in the pivoted wooden doors. The building was also upgraded to provide better indoor thermal conditions and in terms of other services. For example, the waterproofing on the terrace slabs was improved, layered with external thermal insulation, and a layer of white high-reflective ceramic chips was added to the top of the roof. Air cavities of non-structural walls on the south façade that receive direct sunlight throughout the year were filled with thermal insulation, buoyancy drive self-extractor air ventilator fans were installed on the rooftop slab as a temperature and ventilation control strategy. The efforts to address issues of thermal comfort through passive means have yielded some results in saving of operational energy and higher degree of thermal comfort. This chapter, however, does not elaborate on operational energy but keeps the focus on embodied carbon and embodied energy. Considering the focus on embodied energy, this part of the chapter will detail the process of conservation of the primary elements of this 20th-century heritage structure; exposed brick load bearing and partition walls, and form finished RCC slabs, beams, fins and gargoyles.

One of the major challenges for conserving the structural elements was that the original structural drawings or material specifications were unavailable, and thus arriving at new material as well as structural design was entirely dependent on the testing as mentioned earlier. However, site observations, studying the archived communications between the architects and the engineers reflect that the bricks were made at the kiln located at the site, the cement used was 33-grade Ordinary Portland Cement (OPC), sand came from the river bed that is located about 30 km north of the location, aggregates came from the quarries located about 100 km east of the site and the reinforcement used was Fe250 grade, round section plain mild steel bars. Concrete core tests revealed that the density of concrete varied from $2180 kg/m^3$ to $2380 kg/m^3$, which was below the current standard minimum of $2400 kg/m^3$, indicating porosity and permeability in concrete elements, resulting in high water absorption (5.33% to 7.68%). The carbonation in the slabs was seen in some cases up to the full depth of the slab (75 mm) and in some beams the range was from 50 mm to 175 mm from the surface of the concrete. With 75 mm depth of the slab, and very low concrete cover to the reinforcing steel (in some cases 10 mm), the decayed areas of the slab not only had a carbonated concrete cover, in some cases carbonation went through and resulted in corroded reinforcements and an entire section of decayed slab. In other cases, especially beams, the concrete cover was carbonated, exposing the reinforcing steel. Thorough rebound hammer tests on all concrete surfaces were carried out to arrive at the decision about which specific parts of the concrete surfaces needed repairs and which needed to be removed and recast.

On areas where the concrete cover was damaged, patch repair was done using polymer concrete (polymer modified mortar), while adding additional reinforcement where required. For the restoration of entire elements of RCC, mostly top slabs, fins and gargoyles, all elements exposed to weather, a concrete mix design was prepared that was compatible with the old concrete in order to maintain the structural and visual integrity of the building. During August and September 2020, seven different trial mixes for M25 grade concrete were designed and sample cubes prepared using OPC 43 grade (1 mix), Portland Slag Cement (PSC – 2 mixes), Portland Pozzolana cement (1 mix) and a combination of OPC and white cement (3 different proportions) as per IS 456-2000 and IS 10262-2019 to come up with a compatible concrete mix. These sample cubes were tested for their slump value as well as cube compressive strength at 7 days and 28 days after curing. The colour of the concrete cubes was also compared with the original concrete after allowing them to dry for a week post-curing period. Here, the samples using slag cement were found to be the closest match. The average compressive strength of the concrete cubes prepared using slag cement was found to be 32.69 N/mm^2 for mix (1:1.6:3.17) and 32.46 N/mm^2 for (1:1.44:2.9) which indicates successful mix for M25 grade After considering the compressive strength and colour properties it was concluded that the concrete using Portland Slag Cement would be the most appropriate for the restoration work at the site and was hence approved for the work. In addition to being a sustainable alternative to OPC, slag cement is also advantageous in the construction and repair process due to its low heat of hydration. In order to avoid potential detrimental components from water during the time of construction, water samples from five different borewells were tested and the one with balanced readings of pH values, TDS (measured as particles per millilitre) and chloride was used. The existing reinforcement bars were tested for their yield strength, percentage elongation and ultimate strength along with percentage reduction in the cross-sectional area. It was observed that despite the bars being corroded, their mechanical properties were not drastically reduced, and their ductility was in the range as recommended by IS 432. Hence, it was decided to use the same reinforcement and only provide extra where needed. The new reinforcement was of the same brand, same specification and similar properties.

Aspects of visual integrity also required looking into the question of shuttering pattern. The slab constructed in the 1960s used two types of shuttering, rectangular metal panels of approximately 60 cm × 30 cm for the beams, slabs and fins, and 15 cm linear wooden slats for the vertical parapets. Each of these gave a particular kind of texture, scale and colour to the concrete. These shuttering materials were sourced from the old storage of various contractors from the city ensuring that the tectonics of the space were not altered. Applying polymer concrete to the partially damaged concrete surfaces had similar concerns, considering that the visual texture is very different from an RCC surface. While the structural integrity of the polymer repair was achieved by a step-by-step process of removing the existing surface plaster, chipping weak concrete, removing concrete embedded in rusted reinforcement, cleaning the rust on reinforcement, sealing the damaged or honeycombed concrete with injection grouting, inserting mild steel sheer keys, adding additional reinforcement where required, applying alkaline passivating and bonding coat over the cleaned reinforcement, before applying and curing the self-compacting polymer concrete on the surface; the patched surfaces were left to be seen as repairs. This decision was informed by the philosophy of honest repairs and conservation principles widely and internationally accepted (Macdonald and Goncalves, 2020). During the process of conservation, there were instances of having to take decisions that minimally altered the visual integrity of the building. In all such instances, structural integrity, safety and durability were prioritised.

In the case of the brick walls, it was found that the water absorption of the older bricks ranged from 13.91% to 16.23%, the average percentage being marginally lower than the bricks available at the time of conservation, which indicated water absorption range of 15.76% to 18.84%. Both ranges were found to be within the acceptable limit of 20% for the class of bricks they were. Both the old and new bricks did not show any efflorescence, indicating that the efflorescence found on the site was a result of the continuous water seepage from the damaged plumbing and aged waterproofing. The older bricks were also the same standard as available today in the market. The only difference in testing the bricks of the School of Architecture and the bricks found in the market now was that the older bricks had a much higher compressive strength than the standard bricks available at the time of conservation i.e. 2020–21. Of the five samples tested, the older bricks showed the range of compressive strength to be $7.61N/mm^2$ to $11.81N/mm^2$, while the new bricks tested showed a range of $4.28N/mm^2$ to $7.13N/mm^2$. These findings resulted in some key decisions for conservation of load bearing as well as the non-structural partition walls. It was decided that the partition walls were to be carefully removed and replaced owing to their deteriorated condition and the risk posed by the reinforcement within them. The salvaged bricks from this were then used to repair the load bearing structural walls. This ensured that the structural walls continued to have bricks with similar compressive strength. While this was not crucial for replacing certain surface bricks that had deteriorated due to weathering and leakage, this decision was critical to repair the structural crack located in the north-west corner of the building.

Partition walls were reconstructed using newer bricks and were reinforced by using steel cable mesh that would be less prone to corrosion and expansion. Reconstruction also allowed the possibility of introducing thermal insulation that would affect the indoor temperature, especially during days of extreme high temperature. Mortar tests revealed that the brickwork was done using a mortar ratio of 1:6 (1 cement to 6 graded coarse sand). New construction specifications were matched to that to avoid differential behaviour. To address the issue of efflorescence due to water, potable water was used for mortar where the water/cement ratio was specified to be equal to or less than 0.45.

Embodied Carbon in Buildings, an Indian Context

The case of conservation and upgrading of the School of Architecture at the CEPT University indicates a significant opportunity to mitigate embodied emissions from buildings in India by refurnishing, renovating and restoring available building stock, possibly beyond the concerns of heritage. This is more significant when India's building sector has been showing year-by-year growth for the past three decades with half of the buildings needed in the country by 2050 yet to be built. The building industry in India has a large potential to mainstream calculations of embodied energy and embodied carbon to help build an inventory of practice, inform policy, and ensure India's development while achieving national and global climate goals. The case of the School of Architecture is particularly relevant considering that the two primary building materials, cement and steel, are top contributors to embodied emissions coming from building construction activities, as they are responsible for 60% of total emissions (Kansal, 2022). The case indicates that restoring existing buildings is an effective way, especially when the buildings are functional and have a probability to function for the next 60 years or more. A 3000 kg of RCC needs approximately 1000 kg of steel and 1000 kg of cement that leads to approximately 1000 kg of CO_2 emissions. India is the second largest producer of cement (Tiseo, 2024), emitting about 164 million metric tons of CO_2 a year, with 10 to 12% production growth year by year.

The construction sector in India is responsible for the largest share of CO_2 emissions (22% of the total) into the atmosphere. As established earlier, cement and steel (>75 million tons per annum and >10 million tons per annum, respectively) are the largest of them as they are the most used (Reddy and Jagadish, 2003). A 1995 study (Debnath and Singh) also found that bricks (41%), cement (33%) and steel (9%) represent a major portion of the total embodied energy of materials for a single storey house with load bearing walls. For a naturally ventilated building in an Indian context, as it was found in these studies, the embodied energy and the operational energy are 2.45–2.85 GJ/m² and 2.5– 4.05 GJ/m², respectively, during its service life. While this comparison may not be accurate considering the inconsistencies in system boundaries, analytical methods, geographical locations, and data quality, it is interesting to note that in the context of India, they are similar. And thus, addressing the question of embodied energy may be an important part of achieving climate goals. Considering this, India's draft Energy Conservation and Sustainable Building Code (2024) by the Bureau of Energy Efficiency and draft National Building Code by the Bureau of Indian Standards (BIS) have included the embodied energy in building material as a reporting requirement.

Factors for Calculating the Embodied Carbon for the School of Architecture

The remainder of this chapter outlines the life cycle assessment of the School of Architecture Building, constructed in the 1960s and conserved between 2020 and 2021, following the ISO 14044 – see Table 4.1. This is carried out with the intention of testing the hypothesis outlined in the earlier section that conservation or building restoration proves to be less carbon intensive than reconstruction.

In order to do so the LCA considers a system boundary based on the following considerations (diagrams for system boundaries of the building as well as the life cycle stages are outlined in Figures 4.4, 4.5, 4.6 and Table 4.2).

- The study limits itself to the core and shell of the superstructure and focuses on the significant materials, i.e. cement, steel and burnt clay bricks. As explained earlier, these materials not only make up the major quantity of the building, they also have the highest embodied carbon emission factors. Base quantities and restoration quantities are considered only for these materials for

Table 4.1 Mandatory elements and parameters determined for the Life Cycle Impact Assessment of the School of Architecture building as per ISO 14044

Impact Category	Climate Change
Life Cycle Inventory Result	Amount of greenhouse gas emissions per functional unit
Characterization Model	Baseline model of 100 years of the Intergovernmental Panel on Climate Change
Category Indicator	Infrared Radiative Forcing (W/m²)
Characterization Factor	Global Warming Potential at a 100-year baseline (GWP_{100})
Functional Unit	kg of material
Category Indicator Result	kg CO_2 per functional unit for each construction material

(Source: ISO 2006) p. 18, Table 4.1 (CEPT).

the School of Architecture and all other materials as used for the finishes and fenestration are excluded.

- The study assesses carbon impacts of two scenarios – one in which the building is restored (Restoration scenario) and another in which the building is completely reconstructed the way it is (Reconstruction scenario).
- Furniture, such as stools, tables, soft boards and chairs, are outside the scope of calculations, as their use is a function of occupant behaviour and not building performance.
- All elements except the ones forming the building's structure, such as doors, railings, etc. are outside the scope.
- Trees and lawns on the CEPT campus are not considered for any carbon removal impacts considering that the intention is to study the difference between restoration and reconstruction.
- At the time of restoration, the School of Architecture building was retrofitted with air conditioners and roof extractors for enhanced indoor thermal comfort. The carbon impacts arising from this installation and operation are outside the scope of this study.
- Carbon impacts due to construction and installation life cycle stage (A5 in Figures 4.8 and 4.9), from when the building was originally built and when the building was restored, are excluded from the study due to data unavailability.
- Impacts arising from the building's operational energy use and water use have not been calculated.
- Geographical and temporal boundaries for the embodied carbon emission factors data have been addressed through the data quality indicators in the following sections.

In order to ensure compatibility across materials and life cycle stages, all quantities need to be expressed in a common, consistent unit, i.e. the functional unit (fu). For this study, the functional unit is kilograms (kg). This is consistent with the carbon emissions unit, $kgCO_2/kg$. Please note that the burnt clay brick quantities are expressed in cubic

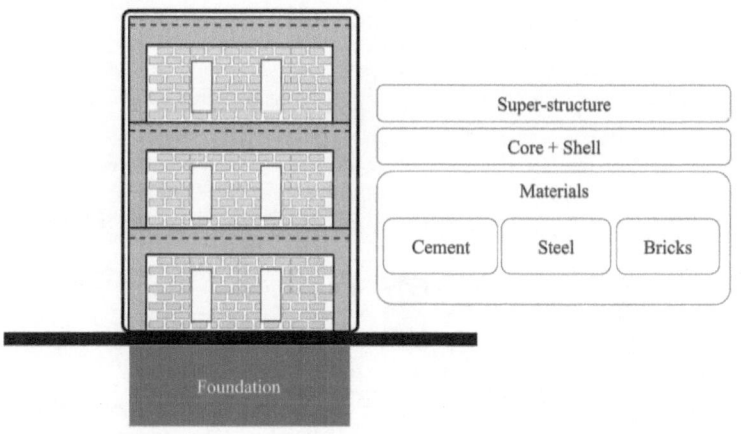

Figure 4.4 Systems and materials covered under this study.
(Sneha Asrani)

Table 4.2 Life cycle stages of a building, and those included in Restoration and Redevelopment scenarios

Life cycle stage			Life cycle stage included in Restoration scenario	Life cycle stage included in Redevelopment scenario
Production	A1	Raw material extraction and procurement	✔	✔
	A2	Transport	✔	✔
	A3	Manufacturing	✔	✔
Construction	A4	Transport	✔	✔
	A5	On-site installation	✖	✖
Use	B1	Use	✖	✖
	B2	Maintenance	✖	✖
	B3	Repair	✔	✖
	B4	Refurbishment	✔	✖
	B5	Replacement	✔	✖
	B6	Operational energy	✖	✖
	B7	Operational water	✖	✖
End of life	C1	Destruction and demolition	✖	✖
	C2	Transport	✔	✔
	C3	Water processing	✖	✖
	C4	Disposal	✖	✖
Beyond lifecycle	D1	Reuse	✖	✖
	D2	Recovery	✖	✖
	D3	Recycling	✔	✔
	D4	Exported energy	✖	✖

(CEPT).

Life cycle stages mentioned in [Table 4.2] have been taken from Figure 6 [p. 21] of EN 15978: 2011 (BSI 2011).

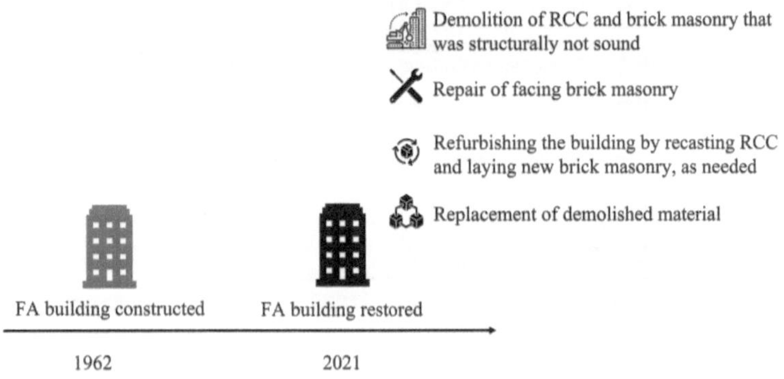

Figure 4.5 System boundary: Life cycle stages covered in Restoration scenario (Base diagram reference: ISO 2006). (Sneha Asrani)

 Demolition of RCC and brick masonry that was structurally not sound

 Repair of facing brick masonry

 Refurbishing the building by recasting RCC and laying new brick masonry, as needed

Replacement of demolished material

FA building constructed　　FA building restored

1962　　　　　　　　2021

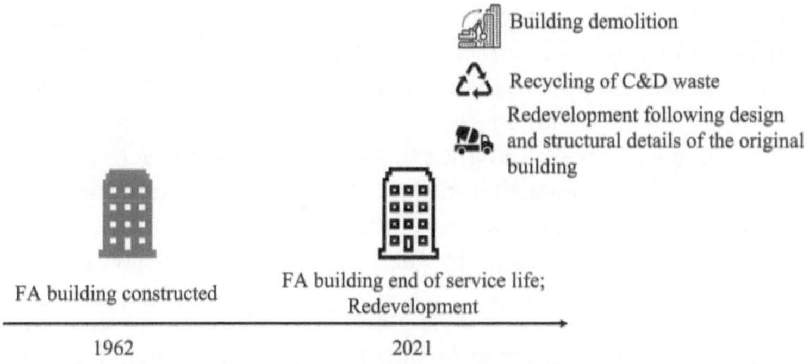

 Building demolition

 Recycling of C&D waste

 Redevelopment following design
and structural details of the original
building

FA building constructed

FA building end of service life;
Redevelopment

1962   2021

Figure 4.6 System
boundary: Life cycle
stages covered in
Redevelopment
scenario (Base diagram
reference: ISO 2006).
(Sneha Asrani)

metres (CuM). This is owing to the general practice of carbon emission factors for burnt clay bricks being expressed as $kgCO_2$/CuM while the density may or may not be reported. This study follows the methodological framework as described in ISO 14044 (ISO, 2006). Adhering to this method, the study has already identified the goal and the scope (system boundary). The next parts of the study include life cycle inventory analysis, life cycle impact assessment and life cycle interpretation. Life cycle inventory analysis requires the study of certain pre-established parameters, such as impact category, inventory result, etc. (please refer to Table 4.1). Considering that this study is aimed at estimating the carbon impact of the building, the study functions within the Climate Impact category (as mentioned in Table 4.1). All other parameters are presented in Table 4.1 and are as per the standards (CSN, 2019). A characterisation factor is applied to the inventory results for conversion to the functional unit. As mentioned above, the characterisation factor will follow the units $kgCO_2$/kg or $kgCO_2$/CuM for bricks. However, where the data for embodied carbon was not available, this value was calculated using embodied energy or energy intensity numbers, expressed in terms of MJ or kWh.

For this study, the carbon emissions for cement, steel and burnt clay bricks are to be studied for two specific time periods, the 1960s, when the building was first constructed and for the 2020s when the it was restored. The emission factors for the 2020s are sourced from literature and follow the functional units as mentioned above. However, no resources were available for the 1960s and they have been estimated using the energy intensity numbers for production of materials of the time, and then by adding the present-day raw material and extraction (A1 in diagram), and transport (A2). Please refer to **Annex A and B** for calculations. ISO 14044 recommends a 'cut off criteria' as a possible way of excluding materials for calculations. In this case, considering that it was possible to calculate the quantities of material, no cut off criteria was applied.

Considering the system boundaries (refer to Figures 4.8 and 4.9) the embodied carbon is calculated here in three stages: Production and Construction (stage A); Use Phase (stage B); End of Life (stage C); and Beyond Life Cycle (stage D). Processes and criteria for data collected for the calculation are outlined here. The as-built bill of quantities for the RCC and brick masonry are available, each of these materials were further broken down into their raw components, i.e. RCC was broken down to cement, sand, aggregate and steel, and brick masonry was broken down to bricks, cement and sand. The emission factors for the raw materials are available in $kgCO_2$ or $KgCO_2e$. Considering that quantities of these materials have maximum carbon impacts, the following answers were sought for the raw materials themselves:

- Market regimes were significantly different at the time of construction of the School of Architecture (1960s), the manufacturing technologies were significantly more energy intensive compared to the ones that are prevalent today. Given this scenario, what would have been the embodied carbon (upfront carbon) footprint then?
- If the exact building has to be constructed now, in the 2020s (when the building was restored), and considering the new, improved technologies, what would be the embodied carbon?
- At the time when the decision of conserving the School of Architecture was made, it had achieved its stipulated service life of 60 years. For this, partial demolition of certain elements and removal of non-structural brick walls was done, RCC was re-casted, new brick masonry was introduced as partition walls and bricks recovered from the existing partition walls were used to repair the damages in the structural walls. How carbon intensive is this process?

Embodied Carbon in the School of Architecture

Stage A: Production and Construction

The current day (2020s) cradle-to-gate or A1–A3 carbon emission factors for cement, steel and bricks have been taken from literature, Table 4.3 mentions the values. These values are specific to the geographical (India), and temporal (values published in 2023) context of the study. They present the kilograms of CO_s emitted due to the extraction, transportation and manufacturing per kg of material produced.

The retrospective (1960s) cradle-to-gate carbon emission factors for cement, steel and bricks are not readily available in literature or open-source databases. A couple of studies (Bapat et al. 2007; Aswale 2015) shed light on the shifting of trends in the cement and brick manufacturing sectors; they mention emission trends between the 1990s and 2020s. However, modelling the retrospective emission factors by extrapolating from that information would have led to a crude estimate with high uncertainty. Thus, the most reliable path to arriving at the retrospective A3 emission factors was to look for the energy use intensity for manufacturing cement, steel and bricks in the 1960s, and convert them to carbon values using furl calorific values. Table 4.2 presents the present day and retrospective cradle-to-gate emission factors.

Table 4.3 Recent and retrospective A1–A3 emission factors for cement, steel and bricks

Material	A1–A3 Emission Factor (present day)	A1–A3 Emission Factor (retrospective)
Cement	0.842# kg CO_2/kg (Kumar 2023) #data represents 2019–2020	1.25 kg CO_2/kg A1 and A2 emission numbers from Chetia et al. (2024), A3 emission derivations to be found in Annex B
Steel	2.5 kg CO_2/kg crude steel (Kumar 2022)	4.58 kg CO_2/kg crude steel A1 and A2 emission numbers from Chetia et al. (2024), A3 emission derivations to be found in Annex B
Bricks	198.87 kg CO_2/cum of brick (Maithel 2023)	215.40 kg CO_2/kg crude steel A1 and A2 emission numbers from Chetia et al. (2024), A3 emission derivations to be found in Annex B

(CEPT).

It is to be noted that carbon emissions–related data considering A1 and A2 life cycle stages for the point in time when the building was constructed was unavailable. Moreover, modelling these numbers would not prove fruitful as the result may not be reliable. As a conservative measure, we have considered the present day A1 and A2 life cycle numbers to be the same.

The bricks used for the School of Architecture building in the 1960s were produced at the site itself. However, location of the plant from where cement and steel were ordered from is unavailable. In that light, it is assumed that the material was ordered from the nearest plant. The nearest manufacturing plant for cement was located in Ranavav, approximately 380 km from the site, and the nearest steel manufacturing plant was located in Bhilai, Chhattisgarh, which is approximately 1100 km from the site. For the present day A4 emissions, the average distances for cement, steel and bricks have been taken from city profile reports (EU-REI and Development Alternatives 2020), which are 50, 200 and 30, respectively.

As far as road transport is concerned, the emission factors have been taken from Gajjar and Sheikh 2015, this study was conducted in 2009 and mentions emissions depending on the type of vehicle. Please refer to Table 4.4 for A4 emission factors.

Table 4.4 Road transport emission factors presently considered for India

Vehicle Category (capacity)	Road transport emission factors (Gajjar and Sheikh 2015) kg CO_2/km
LDV (<3.5 T)	0.3070
MDV (<12 T)	0.5928
HDV (>12 T)	0.7375

(CEPT).

Considering the above limitations and assumptions, Tables 4.5 and 4.6 provide A1 to A4 carbon emissions of primary materials for the construction of the School of Architecture in the 1960s, and in the 2020s at the time of restoration. Due to unavailability of a retrospective emissions factor for road transport, the recent values published in 2010 are used for calculation. Electricity and fuel used at the time of construction of the building, which is crucial to calculating the emissions arising from construction and installation related activities is unavailable. This part of the life cycle assessment (A5) is thus omitted from the study for both scenarios.

Table 4.5 A1–A4 emissions for FA building considering retrospective emission factors for A1–A3 for Restoration scenario

Material	Quantity	Unit	A1–A3 carbon emission factor (1960s)	Resultant A1–A3 emissions (1960s)	Distance between manufacturing plant and CEPT Campus (1960s)	No. of trips needed to transport required material (1960s)	A4 emission factor (1960s[c])*	Resultant A4 emissions (1960s)
			$kg\,CO_2/$ unit	$kg\,CO_2$	km	nos	$kg\,CO_2/$ km	$kg\,CO_2$
Cement	347264.70	kg	1.25	435271.30	380[a]	29	0.7375	8127.75
Steel	57191.44	kg	4.58	261879.61	1760[b]	5	0.7375	6490.00
Bricks	570.62	cum	215.40	122911.93	0	81	0.7375	0

(CEPT).

a   assuming that the cement was ordered from the Ranavav manufacturing plant
b   during the 1960s, the nearest steel manufacturing plant was located in Bhilai, Chhattisgarh
c   due to unavailability of retrospective emission factor for road transport, the recent value, as published in 2010 has been used for calculation.

Table 4.6 A1–A4 emissions for FA building considering recent emission factors for A1–A3 for Redevelopment scenario

Material	Quantity	Unit	A1–A3 carbon emission factor (recent)	Resultant A1–A3 emissions (recent)	Distance between manufacturing plant and CEPT Campus (recent)	No. of trips needed to transport required material (recent)	A4 emission factor (recent)*	Resultant A4 emissions (recent)
			$kg\,CO_2/$ kg	$kg\,CO_2$	km	nos	$kg\,CO_2/$ km	$kg\,CO_2$
Cement	347264.70	kg	0.84	292396.88	250	29	0.7375	5346.875
Steel	57191.44	kg	2.50	142978.61	300	5	0.7375	1106.25
Bricks	570.62	cum	198.87	113479.55	32	81	0.7375	1911.6

(CEPT).

*   Using a vehicle with 12 T capacity.

Stage B: Use Phase

The use phase of the life cycle of a building includes energy consumed for day-to-day use (B1), maintenance (B2), repair (B3), refurbishment (B4), replacement (B5), operational energy (B6) and operational water (B7). The School of Architecture building, till the time it was restored was a naturally ventilated structure. The B1 and B2 emission values would emerge from day-to-day wear and tear of material surfaces and their maintenance. Since the information about this from the time of construction to the time it was restored is sketchy at best, these values are not calculated. Similarly, considering that the objective of the study is to assess the differences in embodied carbon values of restoration and reconstruction, the operational energy (B6) and operational water (B7) are also excluded. In order the calculate the emission values of repair, refurbishment and replacement (B3 to B5), of the three primary materials identified earlier, the study considers the following three activities taken up at the time (between 2020 - 2022).

- The structural walls of this exposed brick building were repaired with joints being raked, pointed and well defined. The emissions occurring due to material, fuel and electricity consumption is calculated towards the repair, life cycle stage (B3).
- Partitions walls of the structure were removed and replaced with new masonry with new bricks.
- Slabs, beams and other reinforced cement concrete (RCC) members that were severely damaged were demolished and recast and the others were repaired with polymer concrete. The new concrete introduced was prepared to match the structural and visual properties of the older concrete. The impacts of these activities are accrued to refurbishment and replacement (B4 and B5) cycles. Tables 4.7 and 4.8 outline these impacts.

As mentioned earlier, carbon impacts arising from the operational energy consumption and water use (B6 and B7) have been excluded from the study's scope. Importantly, B1–B7 life cycle stages are out of scope for the Redevelopment scenario considering that the repair and refurbishment stage does not apply to it.

Table 4.7 B3 – repair – life cycle stage emissions for FA building for Restoration scenario

Material	Quantity	Unit	Emission factor	Resultant emissions
			$kg\,CO_2/kg$	$kg\,CO_2/kg$
Cement	0.24	KG	0.84	0.20

(CEPT).

Table 4.8 B4 and B5 – refurbishment and replacement – life cycle stages emissions for FA building for Restoration scenario

Material	Quantity	Unit	Emission factor	Resultant emissions
			$kg\,CO_2$/unit	$kg\,CO_2$
Cement	112280.24	kg	0.84	94315.40
Steel	13359.30	kg	2.50	33398.25
Bricks	85.76	cum	198.87	17055.95

(CEPT).

Stage C: End of Life

This section includes carbon impacts of demolition of the School of Architecture building if that were to be the case. These calculations help with building the Redevelopment scenario as opposed to the Restoration scenario for which the calculations are already undertaken as part of the B stage. However, the building is not yet demolished. The calculations of these stages, thus, depend on a standard assumption that they are to be estimated at 50% of the construction and installation impacts. Considering that the construction and installation impacts (A4) are not considered for this study, the impacts of the actual activities of demolition (C1) are not considered either. The impacts of transport of the demolished material from the site to the recycling plant have been quantified. If demolished, the material would be sent to the Construction and Demolition (C and D) plant at a distance of 50 km from the CEPT Campus. The total carbon impact in that case is calculated based on the expected weight and the number of trips to the plant. Table 4.9 below presents the transport emissions factor (C2) for material removed at the time of restoration, and Table 4.10 presents a hypothetical case of transport emissions that would incur in case of complete demolition for redevelopment. Please note that the impacts from water processing and waste disposal are not counted here.

Table 4.9 C2 – transport between CEPT Campus and recycling plant – emissions for FA building for Restoration scenario

Material	Quantity	Unit	Distance between manufacturing plant and CEPT Campus (recent)	No. of trips needed to transport required material (recent)	A4 emission factor (recent)*	Resultant A4 emissions (recent)
			km	nos	$kg\,CO_2$/km	$kg\,CO_2$
Cement	33057.49	kg	50	3	0.7375	97.62
Steel	4405.42	kg				13.01

(CEPT).

* Using a vehicle with 12 T capacity.

Table 4.10 C2 – transport between CEPT Campus and recycling plant – emissions for FA building for Redevelopment scenario

Material	Quantity	Unit	Distance between manufacturing plant and CEPT Campus (recent)	No. of trips needed to transport required material (recent)	A4 emission factor (recent)*	Resultant A4 emissions (recent)
			km	nos	kg CO$_2$/km	kg CO$_2$
Cement	347264.70	kg	50	115	0.7375	1074.05
Steel	57191.44	kg				176.89
Bricks	570.62	cum				2989.69

(CEPT).

* Using a vehicle with 12 T capacity.

Stage D: Beyond Life Cycle

The standard practice in India would be to recycle the C and D waste, and the literature on this suggests an emission factor (Joshi et al. 2024). Based on that, Table 4.11 indicates the emission for the Restoration scenario and Table 4.12 indicates the same for the Redevelopment scenario. The quantities considered for the Restoration scenario are based on the quantities of material that were replaced at the time of restoration (mentioned in the earlier section). The Redevelopment scenario, however, considers demolition of the entire structure and the quantities are considered accordingly.

Table 4.11 D3 – recycling – emissions for FA building for Restoration scenario

Material	Quantity	Unit	Emission Factor	Resultant emissions
			kg CO$_2$/kg	kg CO$_2$
Cement	33057.49	kg	7.54E-05	2.49
Steel	4405.42	kg		0.33

(CEPT).

Table 4.12 D3 – recycling – emissions for FA building for Redevelopment scenario

Material	Quantity	Unit	Emission Factor	Resultant emissions
			kg CO$_2$/kg	kg CO$_2$
Cement	347264.70	kg	7.54E-05	26.18
Steel	57191.44	kg		4.31
Bricks	570.62	cum		72.87

(CEPT).

The Life Cycle Assessment (LCA Framework of ISO 14044 (2006) includes conducting data quality checks and analysis, such as sensitivity and uncertainty analysis to reduce discrepancies from data used from multiple sources with varying parameters. For this study, the data is either taken from the actual quantities available about the building and restoration of the School of Architecture, or from published reports such as journal papers that may not be recent but are India specific. The data obtained from literature for this study satisfies the data quality indicators. The material emission values of the 1960s have been calculated using energy intensity numbers. Having said that, only parts of the LCA have been undertaken as a part of this study, and the uncertainty of the emissions thus cannot be absolutely ascertained.

Interpretation

The embodied carbon study has considered two scenarios. The first is the Restoration scenario, as it happened. Carbon emission for that scenario is mentioned in Table 4.13. As per the table, the total emission for stage A (production and construction) can be rounded off to 834,680 $kgCO_2$. The emissions are higher than the stage A emissions for the Redevelopment scenario (refer to Table 4.14), considering that the manufacturing processes were significantly more energy intensive in comparison to the present day. Stage B (use phase) emissions in Restoration scenario would include the actions of repairs taken in 2020 which amount to 144,770 $kgCO_2$ (rounded off). Stages C and D (end of life and beyond life cycle) in this scenario amounts to negligible emission (113.45 $KgCO_2$) considering that the life of the building gets extended. The total carbon emission for the Restoration scenario amounts to more than 979 $MTCO_2$ as shown in Table 4.13.

The LCA of the Redevelopment scenario already has a spent carbon of the existing building. The stage A emissions of the existing building (834,680 $kgCO_2$) are already spent, in this case in the 1960s when the building was constructed. The carbon emissions of the use phase, that would include the operational energy would be added to that, but for the purpose of this study, we have left that outside our scope.

Table 4.13 Life Cycle Carbon assessment for FA building for Restoration scenario

Material	A1–A3	A4	B3	B4+B5	C2	D3
	kg CO$_2$					
Cement	435271.30	8127.25	0.20	94315.40	97.62	2.49
Steel	261879.61	6490.00	0.00	33398.25	13.01	0.33
Bricks	122911.93	0.00	0.00	17055.95	0.00	0.00
Total			979563.34			

(CEPT).

Table 4.14 Life Cycle Carbon assessment for FA building for Redevelopment scenario

Material	$D3_{1960s}$	A1–A3	A4	C2
	$kg\,CO_2$			
Cement	26.18	292396.88	5346.875	1074.05
Steel	4.31	142978.61	1106.25	176.89
Bricks	72.87	113479.55	1911.6	2989.69
Total		561357.02		

(CEPT).

Table 4.14 outlines the carbon emissions of the stages from demolition of the existing building to reconstruction of a new one. The study assumes that the new building would be of a similar nature, as a hypothetical scenario. With the present-day systems of production and construction, the making of the new structure would have 33% reduced stage A emissions (557,220 $kgCO_2$ rounded off) than the emissions when it was built in the 1960s. Total potential emissions for the Redevelopment scenario would be 562 $MTCO_2$ (refer to Table 4.14).

If emissions are perceived as money, the emissions occurring at the time of construction in the 1960s are already spent and any additional carbon emissions, including the hypothetical scenario of redevelopment, must be added to it. Figure 4.7 indicates that the base emission of the School of Architecture building, when it was constructed in the 1960s is 834 $MTCO_2$. The redevelopment scenario would add another 562 $MTCO_2$ to that making the total emissions for materials used as 1396 $MTCO_2$. The Restoration scenario on the other hand adds 145 $MTCO_2$, which is less by 417 $MTCO_2$. Figure 4.8 puts the numbers into a perspective of percentage of emissions and it shows that in case of the Redevelopment scenario, the ratio between the emissions at the time of construction (1960s) and redevelopment (2020s), is 60:40, whereas in case of restoration it is 85:15.

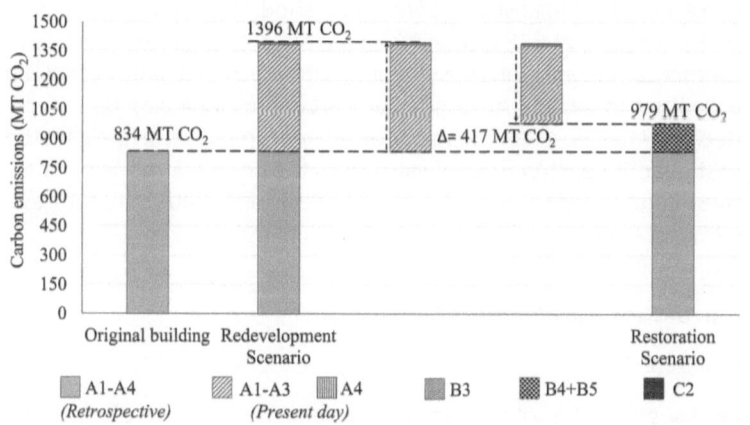

Figure 4.7 Comparative carbon emissions for the material used for the identified scenarios. (Sneha Asrani)

**Figure 4.8 Carbon
emissions for both
scenarios.**
(Sneha Asrani)

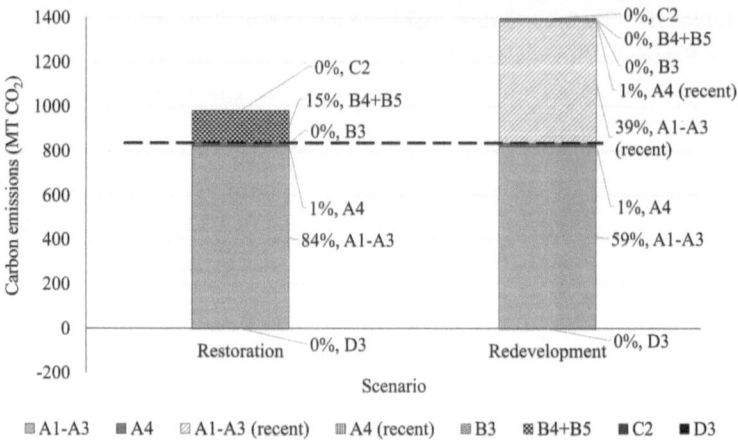

The study here works with the assumption that the operational
energy of a naturally ventilated building, which is now upgraded to be
partially climate controlled, would have similar efficiencies as a newly
constructed building. This assumption is not entirely misplaced considering that the original building is recognised for its design that considers passive design tools such as orientation, thermal mass and
vertical airflows. It is important to state that here because in many
cases older buildings are demolished for their inefficient expenditure
of operational energy.

Last Thoughts

Discussions of economic valuation for cultural heritage, especially in
an Indian or South Asian context, generally take place either in the
space of tourism and visitation, or as an exercise of cost-benefit analysis. The 20th-century heritage of the region, as discussed earlier, are
mostly public architecture or institutions that are still in use and when
they deteriorate, the questions surrounding them are about the cost
benefits. Often, these cost-benefit studies do not consider the environmental costs of emissions already spent at the time of construction. With all the challenges of conserving the concrete structures
constructed in the post-independence period, the study of the School
of Architecture at CEPT, Ahmedabad, India is telling of the long environmental benefits of conservation. The study not only makes an
argument to include the embodied carbon and LCA as a part of the
decision-making matrix, it also gives an opportunity to investigate the
vast quantity of structures of value from the period from the perspective of embodied carbon while making decisions about their futures.

Annex A

Calculating Raw Material Quantities

Concrete

Let's suppose the bill of quantity mentions the cast concrete quantity
to be V cum, taking 1.54 as an expansion factor, the quantity of dry

Table 4.15 Reinforcement steel calculation basis

Element	Reinforcement quantity (as a percentage of concrete volume)
Slab + Beam	2% for First Floor, and 1.0% for Second Floor and Terrace
Walls	2.5%
Fins	1.0%

steel quantity, considering 5% wastage $= 1.05 \times a\% \times 7850 \times V \ kg$

materials as $1.54 \ V$. Now, factoring in 5% wastage, the final quantity of wet concrete is D = 1.617'x'.

For M25 grade concrete, the mix design for cement:sand:aggregate is 1:1:2, thus the quantity of cement has been arrived at using **equation 1**. Please note, we have considered a wastage of 2% in the cement

$$cement \ quantity \left(considering \ 2\% \ wastage\right) = \frac{1(D)}{1+1+2} \times \frac{50}{0.035} kg \quad (4.1)$$

As sand and aggregate are outside the focus of the current study, the calculation for them is not shown here. Now, for steel, we arrived at steel quantity of respective elements using common rules-of-thumb as mentioned in Table 4.15.

Brick Masonry

Similarly, for calculating quantity of bricks and cement from brick masonry quantity, we have followed a similar method, except that the expansion factor for converting dry mortar quantity to wet mortar quantity has been taken as 1.25. This example is for deriving the quantity of facing bricks, having a size of 0.215×0.100×0.067 cum, and a joint thickness of 20 mm

Let's suppose the brick masonry qty is x cum, then

$$Brick \ size = 0.215 \times 0.100 \times 0.067 \ cum$$

$$\therefore brick \ volume \left(BR1\right) = 0.001441 \ cum$$

$$Size \ of \ brick \ with \ mortar \left(BR2\right) = 0.235 \times 0.120 \times 0.087 \ cum$$

$$No. \ of \ bricks = \frac{V}{0.235 \times 0.120 \times 0.087} = N$$

Actual volume of bricks, considering 10% wastage $\left(BR3\right) = 1.1 \ N \times BR1$

Volume of mortar, considering 5% wastage $= \left(V - \frac{BR3}{BR1}\right) \times 1.25 \times 1.05$

The cement calculation can be carried out similar to how it was calculated above for concrete.

Annex B

Calculating Retrospective Emission Factors

[A] Cement

In the 1960s–70s, the wet or semi-wet/dry process was prevalent for the production of OPC. While today only 27% of cement production in India is OPC, at that time OPC would be the norm. Based on our research and understanding, the following shows the calculations for CO_2 emissions for cement manufactured in the 60s and 70s.

Total emissions from cement manufacturing = Emissions from thermal energy production + Emissions from limestone calcination (process emissions) + Emissions from electricity use

Dominant process type (1960–70) = Wet, Semi-wet/dry Process

Thermal energy (from coal) consumption (CII and SSEF 2015; TERI 2014) = 1600 kcal/kg clinker

Electricity consumption (CII and SSEF 2015; TERI 2014) = 130 kWh/ton cement.

Emissions from thermal energy production

Fuel used = Coal

Total thermal energy required (TJ) = $1600 * 4.18 * 10^{-9}$ TJ

Emission factor coal = 94600 $kgCO_2$/TJ

Total emissions from coal burning to produce required heat = $1600 * 4.18 * 10^{-9} * 94600$

  = 0.63 $kgCO_2$/kg clinker

For OPC, taking clinker fraction = 0.95

Total emissions from coal burning to produce required heat = $0.63 * 0.95 = 0.601$ $kgCO_2$/kg cement.

Process emissions due to limestone calcination

Carbon emission factor for process emissions clinker (Eggleston et al. 2006) = 0.52 $kgCO_2$/kg of clinker

For OPC, taking clinker fraction = 0.95

Total emissions = $0.52 * 0.95 = 0.494$ $kgCO_2$/kg clinker.

Emissions from electricity use

Due to unavailability of grid emission factor for the period 1960–70, we assume completely coal-based electricity.

Emission factor coal-based electricity generation (CEA 2006) = 1.1 $kgCO_2$/kWh

Total emissions = (130/1000) * 1.1 = 0.143 $kgCO_2$

Emission factor for A3 life cycle stage = 0.601 + 0.494 + 0.143 = 1.238 $kgCO_2$/kg OPC

Now, adding emission factors from A1 and A2 life cycle stages (Chetia, Mital, and Maithel 2024), the A1–A3 retrospective emission factor for cement = 1.25 $kgCO_2$/kg OPC

[B] Steel

Total emissions from crude steel manufacturing = Emissions originating from the kiln + Emissions from electricity use

Dominant process type = Iron making (Blast furnace), Steel making (Open-hearth process/ Basic oxygen furnace)

Due to limited availability of data, electricity use numbers were not found. Also, thermal energy emissions are predominant.

Data Available

Final thermal energy consumption in 1980 (Schumacher and Sathaye 1998) = 45 GJ/ton crude steel

Final thermal energy consumption in 1995 (Price, Phylipsen, and Worrell 2001) = 37.3 GJ/ton crude steel.

Thermal Emissions

Fuel used = Coal

Emission factor coal = 94600 $kgCO_2$/TJ

Total thermal energy required in 1980 (TJ) = 0.045 TJ

Total emissions (1980) = 0.045 * 94600 = 4.267 $kgCO_2$/kg crude steel

Total thermal energy required in 1995 (TJ) = 0.037 TJ

Total emissions (1995) = 0.037 * 94600 = 3.528 $kgCO_2$/kg crude steel

According to a research paper published in 2021 (Wang et al. 2021):

Emission factor (global) steel production before 1970 = 4.5 $kgCO_2$/kg crude steel

Thereby, looking at the emission numbers for the years 1980 and 1995, and the information available in the article:

Emission factor for A3 life cycle stage. Emission factor steel production (1960–70) = at least 4.5 $kgCO_2$/kg crude steel

Now, adding emission factors from A1 and A2 life cycle stages (Chetia, Mital, and Maithel 2024), the A1–A3 retrospective emission factor for steel = 4.58 $kgCO_2$/kg crude steel.

[C] Solid clay fired bricks

Total emissions solid brick manufacturing = Emissions from the kiln + Emissions from the electricity usage in the process (if applicable)

Predominant kiln type: Fixed Chimney Bull's Trench Kiln

Due to limited availability of data, electricity use numbers were not found (negligible emissions as compared to fuel emissions)

Data available (Majumdar et al., 1968):

Dry ash fuel (form of coal) consumption = 123.675 kg/1000 bricks

Weight of fired brick (assumed) = 3 kg

Calorific value of the coal = 7700 kcal/kg

Energy input for 1000 bricks = (123.675 * 7700 * 4.18)/1000 = 3980.6 MJ

Specific energy consumption (MJ/kg) = 3980.6/(3*1000) = 1.33

Density of fired brick (assumed) = 1650 kg/m³

Specific energy consumption (MJ/m³) = 1.33 * 1650 = 2189.33

Emission factor coal = 94600 kgCO$_2$/TJ = 0.0946 kgCO2/MJ

Total emissions (per m3 bricks) = 2189.33 * 0.0946 = 207.1 kgCO$_2$/m³ solid burnt bricks.

Reference List

Aswale, S. (2015). Brick making in India – History. *International Journal of Financial Services Management, 4*, 11–16.

Bapat J. D., Sabnis S. S., Joshi S. V., & Hazaree C. V. (2007). History of Cement and Concrete in India – A Paradigm Shift. Conference Paper at *American Concrete Institute, Technical Session on History of Concrete*, Atlanta, USA.

BSI (2011). *BS EN 15978:2011 Sustainability of construction works. Assessment of environmental performance of buildings. Calculation method* (p. 21). Sourced from https://www.en-standard.eu/bs-en-15978-2011-sustainability-of-construction-works-assessment-of-environmental-performance-of-buildings-calculation-method/ (last accessed on 14th October 2024).

CEA (2006). *CO2 Baseline Database for the Indian Power Sector Version 1.0* (Issue November). Sourced from https://cea.nic.in/wp-content/uploads/baseline/2020/07/user_guide.pdf (last accessed on 14th October 2024).

Chetia, S., Mital, N., & Maithel, S. (2024). *Product stage carbon emission study of three major construction materials – Cement, steel, and clay fired bricks [manuscript in preparation]*. Sourced from www.GKSPL.in (last accessed on 14th October 2024).

Chatterjee, Malay (1985). 1947–1959: Options after Independence, the evolution of contemporary Indian architecture. In *Architecture in India*, 124–131. Paris and Milan: Electa Moniteur. Originally Publication: Catalogue of exhibition held at Ecole Nationale Superieure des Beaux-Arts de Paris, 1985–86. Sourced from https://architexturez.net/doc/az-cf-123837 (last accessed on 29th October 2024).

CII & SSEF (2015). *Case study booklet on energy efficient technologies in cement industry*. Sourced from https://energy.greenbusinesscentre.com/mv/green cementech/pub24/11.Casestudybooklet_cementsector.pdf (last accessed on 14th October 2024).

Croft, Catherine & Susan Macdonald (eds) (2018). *Concrete, case studies in conservation practice*. Getty Conservation Institute, Getty Publications, Paul Getty Trust, Los Angeles.

CSN (2019). *Sustainability of construction works. Environmental product declarations*. Core Rules for the Product Category of Construction Products (CSN EN 15804+A2). Sourced from https://www.en-standard.eu/csn-en-15804-a2-sustainability-of-construction-works-environmental-product-declarations-core-rules-for-the-product-category-of-construction-products/ (last accessed 6th November 2024).

Curtis, William J. R. (1987). Modernism and the search for Indian identity. *Architectural Review*, August 1887. Sourced from https://www.architec tural-review.com/places/india/modernism-and-the-search-for-indian-identity (last accessed on 29th October 2024).

Eggleston, H. S., Buendia, L., Miwa, K., Ngara, T., & Tanabe, K. (2006). *2006 IPCC Guidelines for National Greenhouse Gas Inventories*. UNEP. Sourced from http://www.ipcc-nggip.iges.or.jp/public/2006gl/index.htm (last accessed on 5th November 2024).

EU-REI & Development Alternatives (2020). *Resource flows in Indian cities: City profile of the construction sector in Ahmedabad*. Sourced from https://www.tara.in/assets/pdf/Waste-Management/Resource-Flows-in-Indian-Cities.pdf (last accessed on 5th November 2024).

Gajjar, C., & Sheikh, A. (2015). *India Specific Road Transport Emission Factors (Version 1.0)* (p. 36). India GHG Program. Sourced from https://shakti foundation.in/wp-content/uploads/2017/06/WRI-2015-India-Specific-Road-Transport-Emission-Factors.pdf (last accessed on 14th October 2024).

ISO (2006). *Environmental management – Life cycle assessment – Requirements and guidelines*. (ISO Standard No. 14044:2006). Sourced from https://www.iso.org/standard/38498.html (last accessed on 14th October 2024).

Joshi, S., Monani, D., & Sahu, A. (2024). Life cycle analysis of the recycling process for construction & demolition waste management: A study of Noida, Uttar Pradesh. *Journal of Applied Science, Innovation & Technology (JASIT)*, 3(1), 10–19.

Kansal, Afsha S. B. (2022). *Decarbonising India's building construction through cement demand optimisation: Technology and policy roadmap*. Alliance for Energy Efficient Economy (AEEE). Sourced from https://aeee.in/our-publications/decarbonizing-indias-building-construction-through-cement-demand-optimization-technology-and-policy-roadmap/ (last accessed on 4th November 2024).

Kumar, P. (2022). *Decarbonizing India: Iron and steel sector*. Sourced from www.cseindia.org (last accessed on 14th October 2024).

Kumar, P. (2023). *Decarbonizing India: Cement sector*. Sourced from www.cseindia.org (last accessed on 14th November 2024).

Macdonald, S. & Goncalves, A. P. A. (2020). *Conservation principles for concrete of cultural significance*. Getty Conservation Institute, Paul Getty Trust, Los Angeles.

Maithel, S. (2023). Improved burnt clay brick masonry: Lowering upfront embodied carbon, improving thermal comfort and climate resilience of new housing in the Indo-Gangetic Plains. *CATE 2023 Proceedings*, 420–429. https://doi.org/10.62744/CATE.45273.1190-452-461

Majumdar, N. C., Ahmad, F. U., & Das, K. (1968). *Survey of Thermal Efficiency of Bull's Trench Kilns. Transactions of the Indian Ceramic Society, 27(1)*, 79–88. Sourced from https://doi.org/10.1080/0371750X.1968.10855635 (last accessed 14th October 2024).

Price, L., Phylipsen, D., & Worrell, E. (2001). *Energy use and carbon dioxide emissions in the steel sector in key developing countries*. University of California, April 2001. Sourced from https://www.osti.gov/servlets/purl/783473-fcGKaj/webviewable/ (last accessed on 14th October 2024).

Reddy, Venkatarama B.V., & Jagadish, K. S. (2003). Embodied energy of common and alternative building materials and technologies. *Energy and Buildings* (open access), *35*, 129–137.

Schumacher, K., & Sathaye, J. (1998). *India's iron and steel industry: Productivity, energy efficiency and carbon emissions*. U.S. Department of Energy Office of Scientific and Technical Information, October, 1998. Sourced from https://www.osti.gov/biblio/753016 (last accessed on 5th November 2024).

Scriver, P., & Srivastava, A. (2016). *India: Modern architectures in history*. Reaktion Books.

Shah, R. J., & Shah, M. (2018). *Preliminary structural assessment of the Faculty of Architecture Building*, Unpublished report submitted to CEPT University, available for reference at CEPT Archives, Ahmedabad.

TERI (2014). Energy demand. *TERI Energy & Environment Data Directory and Yearbook 2013/14* (pp. 208–219). TERI.

Tiseo, I. (2024). *CO_2 emissions from fossil fuel and industrial purposes in India 1970-2023*. Statista. Sourced from https://www.statista.com/statistics/486019/co2-emissions-india-fossil-fuel-and-industrial-purposes/ (last accessed 4th November 2024).

Wang, P., Ryberg, M., Yang, Y., Feng, K., Kara, S., Hauschild, M., & Chen, W.-Q. (2021). Efficiency stagnation in global steel production urges joint supply- and demand-side mitigation efforts. *Nature Communications, 12(1)*, 2066. Sourced from https://doi.org/10.1038/s41467-021-22245-6 (last accessed on 14th October 2024).

5

CARNEGIE LIBRARIES OF BRITAIN

Assets or Liabilities? Managing Altering Agendas of Energy Efficiency for Early 20th-Century Heritage

Oriel Prizeman, Mahdi Boughanmi and Camilla Pezzica

Introduction

Within the 2030 United Nations (2015) sustainable development goals for cities, both SDG 11.7 "universal access to safe, inclusive and accessible, green and public spaces" as well as 11.4 "strengthen efforts to protect and safeguard the world's cultural and natural heritage" are immediately relevant to the maintenance of public library buildings. With respect to social sustainability, Klinenberg (2018) has recently made the assertion that public infrastructure, in the form of public buildings is instrumental in confronting inequality. The Chartered Institute of Public Finance and Accountancy (2019) reported in 2018/19 that there had been a 29.6% drop in spending on public libraries since austerity measures in the UK began in 2009/10. Although it notes a small upturn in the last year, impacts on the least advantaged in society during the Covid-19 pandemic have highlighted inequalities in Britain (Horton, 2020; Paton, 2020), it seems inevitable that the pressure of reduced public finances will be even further compounded by increasing demand for social benefits of public libraries in the near future.

Questioning a values-based approach to sustainable preservation, Avrami (2016) makes a critical point, citing Hobsbawm's (2012) notion of the re-creation of heritage values. She argues that the defence of heritage as a non-renewable resource is countered by its very capacity for the application of values attributed to it to be renewed. A critical risk remains that ambitions for meeting one political agenda should not eclipse those of another. The Carnegie mission was to encourage local engagement with asset management, his gifts were contingent upon councils adopting the Free Libraries Act which enabled them to levy rates in support of maintaining the libraries. Today however,

DOI: 10.4324/9781003527404-7

following library closures, a number of libraries are community managed. In 2018–19 it was reported that cuts to library funding were forcing libraries to rely on 51,478 volunteer workers (*Chartered Institute of Public Finance and Accountancy*, 2019). The trend towards community-led regeneration is increasing. In this context of constrained resources running in parallel with increasing levels of heritage attribution and pressures to mitigate the impact of climate change, justifying the financial cost of operational energy use will often come ahead of anxiety over excessive carbon emissions.

There are 336 surviving buildings in Britain that were funded under Andrew Carnegie's library buildings programme illustrated in Figure 5.1, predominantly designed and built in a short space of time prior to the First World War. The accelerated standardisation of their delivery creates a specific opportunity for considering their collective endurance and consequent indicators of their sustainability retrospectively as noted in previous work (Prizeman, 2012; Prizeman et al., 2020). These buildings are socially important in that they were at the vanguard of delivering newly devised ambitions for "open access" to all, they were also designed with a consciousness of standards evolving transatlantically.

Public libraries are an intensely designated building type, with spatial arrangements as well as fixtures and fittings that must withstand operation by users and at the same time remain manageable within limited budgets. Their continued performance demands in terms of delivering the public with means for navigation and accessibility and managers with capacity for surveillance and energy efficiency pose significant challenges and has resulted in changing attitudes towards their fitness for purpose but also their heritage value. Significant constraints on public funding for libraries make these concerns more acute.

Interrogating our recent complete survey of Carnegie funded library buildings in the UK, we aim here to consider the potential for reasoning critically with respect to reflections on life cycle from the particular to the general and vice versa with an aim of enhancing decision-making tools for conservation management that should be relevant to these buildings and many more of the same era and/or typology. Using historical analysis and current operational data, the research draws out deeper readings of shifting socio-economic agendas that have a bearing on how and why quantitative measures alone might be inadequate tools of assessment. The key feature of this work is to highlight the interplay of changing imperatives both of environmental resources and cultural heritage values leveraged by socio-economic weightings. The challenge of balancing these often-conflicting demands requires us to consider how we can best develop rules-of-thumb or principles that can assist stakeholders, managers and architects.

Whereas modelled projections tend to underestimate actual energy use for new buildings, operational data for 107 public library buildings here suggests the reverse to be true for a consistent cohort of buildings.

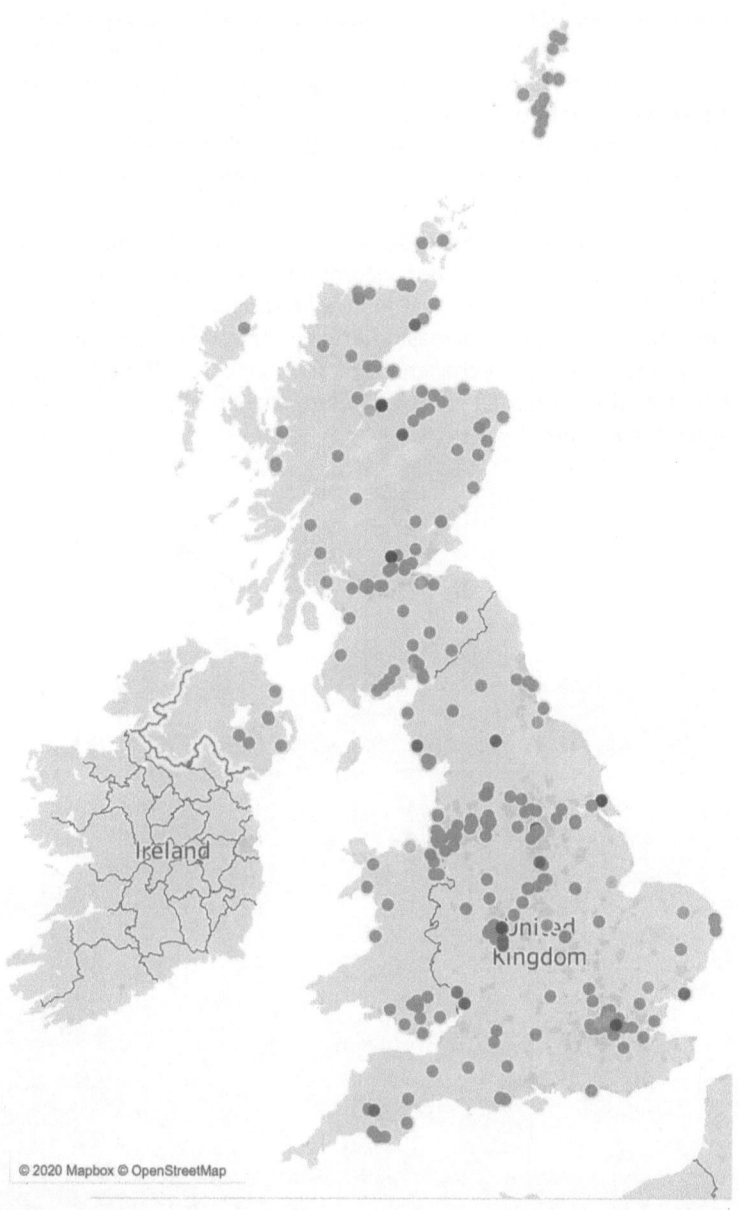

© 2020 Mapbox © OpenStreetMap

Figure 5.1 Map showing location of all surviving buildings funded with a Carnegie library grant in the UK.
(Oriel Prizeman)

Originally built to maximise daylighting for economic reasons, corre-lating data here suggests this characteristic may still contribute to reduced electricity use in these buildings. Reviewing these findings against contextual measures of deprivation further supports their cur-rent value as critical contributors to welfare, providing accessible and amenable public interiors.

The key mission for Carnegie library buildings of responding to an economic context of cheap heat and expensive light has been reversed by the advent of cheap and relatively low energy lights and increased concern with respect to climate change for lower carbon emissions. To some extent, therefore, these buildings were all designed to achieve precisely the opposite conditions that current designers would emulate. In addition, their two-tier single-glazed rooflights are the cause of significant heat loss and vulnerable to leaking. These features are at once both delightful and problematic. With libraries and councils short of funds it is not uncommon on wet days to see buckets collecting rainwater inside and to hear librarians complaining of their need to clear gutters, increasing a sense of vulnerability. Given the iteration of these same challenging conditions across the country, it is critical to draw together evidence in support of best practice for managing their future.

Figure 5.2 Walthamstow central library exterior.
(Oriel Prizeman)

The apparent opposing measures of performance for Walthamstow central library, Figures 5.2 and 5.3, provides a good example of conflicting agendas. Valued for its success as a library and as a heritage asset, ostensibly it under-performs in terms of energy use. In 2019 Walthamstow ranked 13th of all the 3583 public libraries in the UK for the number of visits it supported (*Chartered Institute of Public Finance*

Figure 5.3 Walthamstow central library interior.
(Oriel Prizeman)

and Accountancy, 2019). However, in terms of its display energy certificate it performs badly and is F-rated. Originally built in 1894, its extension pictured here was funded by Carnegie in 1909, both were designed by J. Williams Dunford. The building was given early Grade II listed status in 1973.

Hong (2015) has argued that the benchmarking of operational energy ratings by building type should be reviewed. Here, the implications of benchmarking are revealed against a substantive dataset, raising some further questions for publicly accessible heritage buildings that are also obliged to meet other pressing challenges. Arguably 126-year-old Walthamstow central library's impressive ranking, providing members of the public to make 600,393 visits last year in an area with an Index of Multiple Deprivation decile of 3 (Ministry of Housing, 2019) should qualify to eclipse this measure of performance and is a greater determinant of its sustainability. However, the imperative to reduce carbon emissions by improving energy efficiency in buildings is unavoidable, balancing such demands is not a trivial task. The prevalence of this issue shared amongst a large number of closely matched circumstances determines the need for guidance.

Life cycle prediction methods variously suggest adopting an anticipated range of 50–100 years (Bull, 1993, 2015; Caplehorn, 2011;

Ellingham, 2006, 2013; Flanagan, 1989, 2004; Langston, 2005; Mohamed Abdelhalim & Abouzid Sameh, 2011). The majority of buildings that are still functioning as libraries here are 110+ years old suggesting that they offer an opportunity for life cycle review that acknowledges changing imperatives at scale. With built heritage that is significantly protected, the threshold for determining the viability of alteration is high, however where built heritage has a lower or emerging designated value, the risk of inappropriate alteration or wastage is increased. The transition of a building from being perceived as a liability to that of an asset is influenced by various factors. A key component of this is the abandonment of old buildings justified by anticipated energy reduction. Technical director of CIBSE, Hywel Davies noted in 2013 "The performance of low energy designs is often little better, and sometimes worse, than that of an older building they have replaced, or supplemented" (Cheshire. D, 2013). It is essential therefore to seek means to balance the challenges of both maintaining socio-economic and heritage value whilst reducing carbon emissions through energy use and to better understand what we already have. Here, a range of evidence illustrates how intersecting values of changing socio-economic contexts and cultural sentiment over that last century have weighted political decisions impacting the extent of permissible physical adaptation of buildings in order to achieve greater energy efficiency. Official data compiling the measured energy use of 107 buildings is correlated here with measures of both heritage value and socio-economic context. Modelled estimations of energy use are also collated, inconsistencies in these underlining the influence and risk of pre-judgement with regard to the presumed energy performance of early 20th-century buildings.

Towards Life Cycle Indicators: Research Background

Appealing to instinctive sentiment for historic buildings amongst preservation architects, Elefante (2007) argued that "the Greenest building is ... one that is already built". Theoretically, a dense retrospective analysis of assets could provide valuable indicators for such decisions and was one intention of this research. The aim being to determine in principle that replacement was more wasteful than re-use or adaptation. However, it became evident through making a complete photographic survey of the Carnegie funded library buildings in Britain that more importantly than collating a historic database of embodied carbon (a somewhat ironic task for auditing the philanthropic legacy of a steel magnate in any event), collating evidence of how political and economic agendas have shifted is more informative in terms of qualifying the sustainable potential and future treatment of these buildings.

Statistical outcomes of performance can be rendered meaningless if they are not properly qualified by nuanced contextual readings of design or operational intent. Janda's (2011) paper "Buildings don't use energy, people do" qualified how energy efficiency should be addressed for traditional buildings. There is an expectation that measured energy use in new buildings will generally exceed the modelled prediction. However, here, with a significant and cogent dataset of older buildings it is possible to draw together data indicating distinct measured trends in energy use and to hold these against modelled expectations and challenge this assumption. The aim is that these data provide a platform for decisions in principle.

Life cycle analysis (LCA) tools aim to enable decision makers to balance future outcomes against calculable risks when designing new buildings or deciding to adapt old ones. Data supporting observations of general practice and LCA of common components as opposed to charting complete data for individual buildings emerged as a more relevant tool. There is a frustration when attempting LCA for heritage buildings that the "sunk costs" (Brealey, 2020; Flanagan, 2004) of embodied energy and carbon are discounted. A report by Historic Scotland asserts that for historic buildings, the embodied carbon of the past has no mitigating impact on future consumption (Menzies, 2011). However, in the case of a large number of existing assets that may or may not be protected by heritage legislation and might be replaced with new buildings, finding some means to benchmark the comparative embodied energy between existing and new construction is still a relevant consideration as highlighted in Historic England's (2020) recent report. Dixit et al. (2010) highlighted that there were inconsistencies between parameters used for calculating embodied energy but noted that geographical distance of materials was the measure most universally incorporated in calculations. Therefore, although other parameters for the production of historic building materials, for example attempting to calculate the CO_2 emissions of horse-drawn barge transport, would take some time and have little generalisable value, accumulating data demonstrating the simple measure of distance for principle building materials from cradle-to-gate is readily observed precisely from reports in the contemporary press and more generically from visual analysis of photographs and is therefore presented here.

Addressing relatively modern or marginal heritage, Berg et al. (2018) have demonstrated the validity of life cycle analysis tools for evaluating refurbishment options for a single building modern heritage. Several life cycle research studies using both LCA and life cycle costing (LCC) argue that it is the operational life of buildings, as opposed to their construction that is most responsible for their environmental impact in terms of energy use with a proportion cited of up to 80% (Avrami, 2016; Cole & Kernan, 1996; Mudgal, 2009). The 2007 United Nations Environment Programme report (Huovila & United Nations Environment Programme, 2007) cited in Smith et al. (1998) stating that 50% of buildings' contribution to CO_2 emissions are accounted for by operational costs. It is argued that a combination of passive and active transformations to the supply and use of energy could have a significant impact on subsequent performance (Ramesh et al., 2010).

However, Ibn-Mohammed et al. (2013) have tabulated a range of cited proportions for embodied versus operational emissions which in the UK alone vary from a maximum 80% of life cycle carbon being embodied carbon (Smith & Fieldsin, 2008) to an estimation of embodied energy as between 3% and 35% of 100-year life cycle energy demand. They observe the complexity of comparing varied computational methods, however, the cautionary note is important as it indicates that the assertion highlighting the relevance of operational energy use can be misleading. Indeed, they go on to argue using cited research commissioned by British Land that were the grid to be de-carbonised to 0.2 $kgCO_2e/kWh$ as anticipated by 2030 (Committee on Climate Change, 2008; Hammond et al., 2011), the balance in estimation of embodied carbon for a building currently calculated at 42:58 could shift to 68:32 (Battle, 2010). This anticipated shift of emphasis is relevant to considering how a substantial existing "state" such as that of these buildings, should be assessed in the interim.

Critically, the cost imperatives of efficiency for light and heat have reversed during the last 120 years. The first Carnegie library to open in America at Braddock recorded in a 1916 annual report that the cost of heat was USD 225, 16% whereas that of light was USD 1200, or 84% (Taylor, 1916). Today, the normal expectation for the cost of heat and artificial light is likely to be the reverse. Nevertheless, CIBSE still caution the underestimation of operational energy that is dedicated to artificial lighting (Cheshire. D, 2013). The CIBSE 2012 benchmarks note that for lighting, public libraries, under category 8, good practice is 5 w/m² but typical practice is 9 w/m² (Ed. Butcher, 2012). Research in China derived from 54 surveys determined that library buildings without air conditioning built before 1990 that were under 10,000m² and benefitted from natural lighting, tended to use less energy, around 40 kWh per m² per year, than later buildings over 20,000m² with mechanical ventilation and dependence on artificial lighting in the day which averaged 70kWh per m² per year (Xuan, 2011). The majority of libraries here are heated with natural gas and naturally ventilated. During the last century energy sources have already been changed at least once. Interchanging the source of energy can be a relatively minimal intervention for a listed building. As has been argued by those highlighting the dominance of operational costs (Ramesh et al., 2010) as well as those highlighting the dominance of embodied carbon (Ibn-Mohammed et al., 2013), both note the potential to reduce operational costs through improved technologies used in appliances as well as in the CO_2 of externally supplied energy. In the case of a significantly large group of existing buildings, these indicators are important. The relative impact of altering the energy source is therefore an aspect of LCA that demands consideration here.

Key Characteristics

The strictly prescriptive demands of the Carnegie library design briefs, combined with the significant rapidity of the proliferation of buildings, led to a high degree of standardisation as opposed to a wide variety of innovation in design. This is useful in terms of enabling the valid comparison of similar or even identical attributes with respect to differing contextual conditions. The standards are nuanced in so far as the buildings were generally designed by local architects, using local materials to deliver global standards. They achieved this by referring to detailed design briefs and clear, albeit limited, technical guidance through a small number of authoritative sources and constant reporting in the architectural press. Ironically, although pre-dating stripped back modernist buildings that aesthetically removed tradition from the buildings of the mid-20th-century, the Carnegie aspiration had been "TO OBTAIN FOR THE MONEY THE UTMOST AMOUNT OF EFFECTIV ACCOMMODATION, CONSISTENT

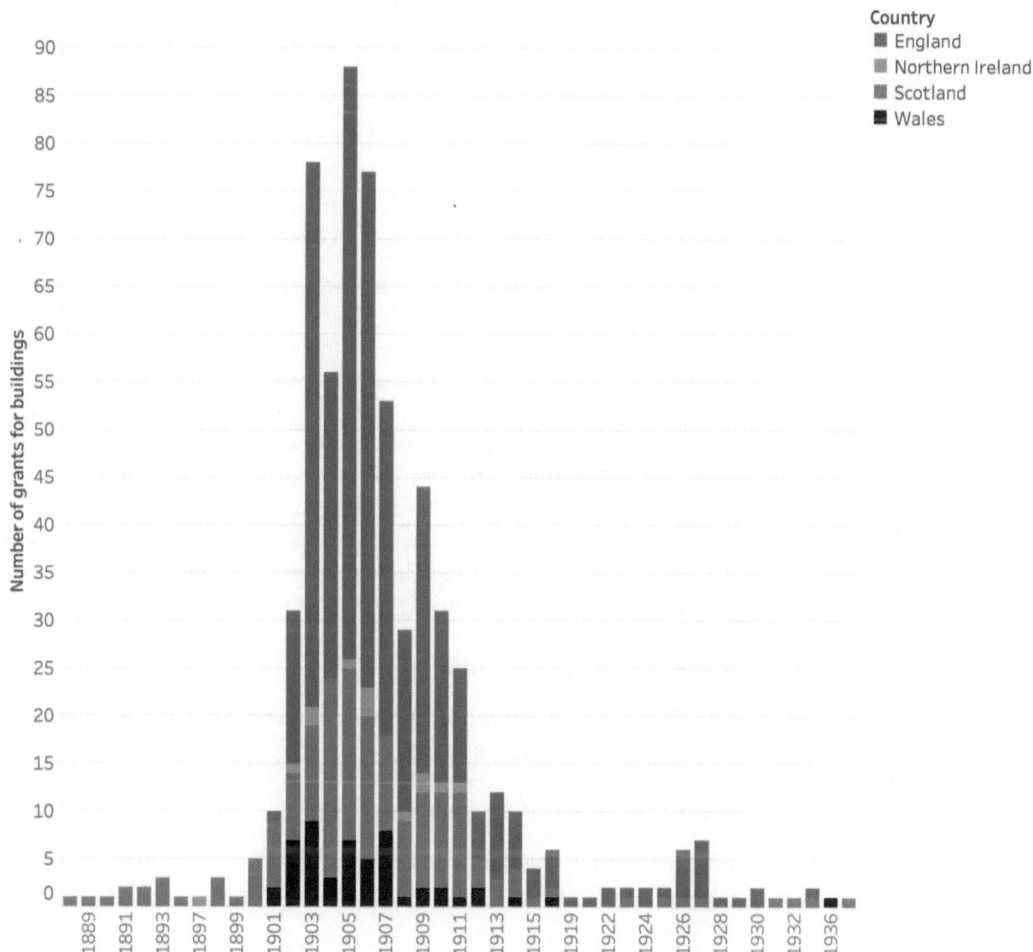

Figure 5.4 Histogram showing dates of all buildings funded with a Carnegie library grant in the UK, shaded by county.
(Oriel Prizeman)

WITH GOOD TASTE IN BILDING" [sic] (Bertram, 1911). As has been discussed in related previous work (Prizeman, 2013), these aspirations are often obscured by the efforts of local architects to aggrandise the appearance of their towns and cities.

The legacy of Carnegie library buildings in Britain is a significant anomaly. His philanthropy transformed the public library movement in Britain, yet it happened at such a pace that its assimilation is hard to distinguish from the Edwardian public domain as a whole. However, today we are used to the notion of chain stores, of global brands and of generic expectations of the basic constituents of the high street. The average lifespan including all buildings funded with Carnegie library grants in the UK has been 98 years. Our survey has determined that of 490 buildings built, 336 survive, 224 of which are open public libraries. Considering how quickly they arrived as illustrated in Figure 5.4 and how many other models have followed suit, it is critical to reflect upon how this substantial inheritance will survive or evolve and adapt. As these buildings become more widely acknowledged as a part of our heritage and are protected as such, they also become a part of the problem with respect to energy efficiency. This reflection is particularly important with respect to the similarities and common features of these buildings as there is an opportunity for measures of energy efficiency to be considered at scale and therefore to have a significant impact.

Brokering the relationship between emergent heritage value and its negotiation with imperatives of reducing CO_2 emissions is difficult, adding the weighting of these building's ongoing merits in serving socio-economic needs compounds this complexity further. However, considering that we can see that these standardised buildings have been re-iterated in waves by so many subsequent 20th-century buildings from fast food restaurants to cinemas, demands that such approaches are addressed. This chapter sets out to quantify and qualify data surrounding the energy use of these buildings and to present it in such a way that these values may be considered against more nuanced values of social benefit and heritage value.

Materials and Methods

The results are structured in three parts each using different datasets and varied methods of analysis:

Identifying Fixed Vulnerabilities

Global and local knowledge transfer
Global components
Local materials

Identifying Changing Agendas

Changing demands: Transformation of socio-economic contexts
Changing role: Heritage value and audit of re-use

Changing estimations of performance: Measures of efficiency

Drawing Life Cycle Indicators

LCA: Toxteth library
Suggested adjustments

Identifying Fixed Vulnerabilities

The data derived from a recent survey is first presented in order to describe common "fixed" vulnerabilities and characteristics of the library buildings. These characteristics are tied to historical circumstance, original design imperatives which have been analysed in previous work (Prizeman, 2012; Prizeman et al., 2020) including a charting of their common features for HBIM.

Survey

As a whole the Shelf-Life project has included the first complete photographic survey of Carnegie library buildings in the UK (Prizeman, 2020). The data supporting the collation of the list of library buildings are various and previously incomplete. In the main, the card index of the Carnegie Corporation of New York's (1898) archive at Columbia University provides a list of grants offered, however, many of the grants are for a number of buildings and private grants are not included. The gazetteer of public library buildings in Britain included in Black et al.'s "Books Buildings and Social Engineering" (2009) together with Brendan Grimes' (1998) book on Irish Carnegie libraries which covers Northern Ireland together with the directories of the Carnegie UK Trust archives held in the National Records of Scotland are the principal sources. For verifying data on the location of architects, the RIBA *Directory of British architects 1834–1914* (Brodie et al., 2001), the various Pevsner guides, the Historic Scotland *Dictionary of Scottish architects* (Walker, 2016) were used. Retrieval of online journal sources for *The Builder* and the *Building News* together with an archival search for all journals that are not online at Cambridge University Libraries and the RIBA Library in London assisted with identifying the architects of the more substantial buildings, however, smaller and more rural grants required historic local media that is now accessible online.

Determining what did not happen or when buildings closed was more difficult than finding what remained, this involved searching every missing library grant in national and local press through the British Newspaper Archive, Welsh Newspapers Online and for some, specific archival searches at the Carnegie UK Trust archive in the National Records of Scotland. Geographic locations and postcodes were verified through Google Earth™, and where possible on library websites. Edina Digimap® Historic Roam Ordnance Survey maps were used to verify the age of the buildings and to generate latitude and longitude.

Identifying Changing Agendas and Measures of Performance

Second, we trace the changing imperatives that have impacted the survival of the buildings and their performance. These data are collated from a range of sources and are variously correlated using geographic and numerical means. Interpretation also demands discussion and qualification to temper the bald conclusions of numerical results, using data visualisation software, Tableau Desktop™, it has been possible to illustrate trends.

Drawing life Cycle Indicators

Finally, constituent elements that contribute to generic reasoning regarding an attempt towards life cycle analysis are evaluated. A model of Toxteth library and parametric model represented by a 3D pdf was built by Mahdi Boughanmi in REVIT™ based on a laser scan survey by Camilla Pezzica and Giovanni Bruschi made using a FARO focus X130 3D laser scanner and registered using Faro Scene™ software. Using a REVIT™ plug-in to upload the take-off of elements, this data was exported to Oneclick LCA™.

Notes on Data Sources

Socio-economic Contexts

The data include separate socio-economic measures drawn from the various Government data Indices of Mass Deprivation (IMD) for England, Wales (WIMD), Scotland (SIMD) and Northern Ireland. Postcodes of remaining buildings were correlated to lower super output areas separately for Scotland, Northern Ireland, England and Wales. IMD are used to target funding for small geographic areas. Whilst Indices of Mass Deprivation across the devolved nations in Britain do follow similar methodologies, they cannot be directly compared (ONS, 2010, revised 2013). The weightings for different domains of deprivation and timescales are not the same. It is therefore only possible to trace comparative measures using postcode lookups within each of England (Ministry of Housing, 2019), Wales (Welsh Government, 2020b; Welsh Government, 2020a), Northern Ireland (Northern Ireland Statistics and Research Agency, 2017) and Scotland (Scottish Government, 28 January 2020) independently. It is important to recognise that within each area, these are not absolute measures but relative ones. Also, as lower super output areas (LSOAs) may only relate to very small areas (e.g. the average population of an LSOA in Wales being 1600), it should be borne in mind that the area which a library serves may cover a range of adjacent but different IMD scores. Nevertheless, the data is valuable in providing metrics to locate the resilient assets in terms of their indicative socio-economic contexts. The data all contain public sector information licensed under the Open Government Licence v3.0.[1]

[1] https://www.nation alarchives.gov.uk/doc/open-government-license/version/3/.

Heritage Value and Re-use

An audit of listing and heritage status was derived from the current websites of Historic England (Historic England), Historic Environment Scotland (Historic Environment Scotland), Cadw (Cadw) and Northern Ireland's buildings database (Department for Communities). The various listing designations of each nation are different, however, in order to ease visual assimilation of data they have been colour-coded from red (highest) to yellow (lowest) to indicate the degree of listing.

Measures of Efficiency

For collating measures of efficiency, Government Display Energy Certificates were collated manually through individual postcode lookups for each open library building in England, Wales and Northern Ireland and Energy Performance Certificates were retrieved by building type where available. As a result of Article 7 of the European Directive on the Energy Performance of Buildings 2003, amended in 2010, 2013 and 2015, it has been mandatory since October 2008 for Display Energy Certificates lasting one year for buildings over 1000 m^2 and for 10 years for buildings between 250 and 1000 m^2 to be displayed for buildings partially occupied by a public authority in England, Wales and Northern Ireland, now requiring them for buildings with a "useful floor area" of 250 m^2 and above. In Scotland, Energy Performance Certificates are used instead as are all UK buildings which are to be let, hence some of the community managed libraries are also included in the dataset. The certificates were individually downloaded from postcode searches on the Scottish EPC register (Energy Saving Trust), the Ministry of Housing, Communities and Local Government Energy Performance of Buildings Data England and Wales provides access to both EPCs and DECs (Ministry of Housing, 2020a) and The Department of Finance Northern Ireland Non-Domestic Energy Performance Register (Department of Finance, 2020). The most recent certificates up to 31 March 2020 were downloaded. Certificates were not available for every open library online and this is noted in the text. Data for water usage is not readily available and, as it is unlikely to be a significant aspect of library use, a benchmark was adopted. The UK watermark programme is cited by Arpke et al. (2005) giving a benchmark for water usage in a public building for libraries that is 0.203 kl/m^2, best practice is 0.128 kl/m^2 (Mudgal, 2009). The benchmark figure is used here as a presumed level of consumption.

Results

Fixed Vulnerabilities

Rapid Standardisation: Global and Local Knowledge Transfer

As noted above, the rapidity of the arrival of Carnegie library buildings in Britain was remarkable. In just four years between 1903 and 1907, 268 buildings were built. The rapid taper of their inception and of their demise at the start of the First World War compresses the potential for a reflective or iterative approach to their design. The news of the philanthropic opportunity spread very rapidly in the national press and within many local newspapers also. In parallel, the professional press, in the form of trade journals for both architects and librarians were chasing the phenomenon. This reporting effectively spread the common and technical understanding of what was to be expected of a contemporary public library building at an urgent pace. It is important because it enables us to plot the pace of "knowledge transfer". At the same time, the tendency of Carnegie and his private secretary, James Bertram, to intervene and comment on design decisions became more intense.

A database of all the Carnegie library buildings connects each to its architect or designer. Although some councils such as Manchester and Liverpool used the same architect to design a set of libraries each with the same plan but a completely different appearance, the majority of architects only designed one Carnegie library. Hence it is accurate to determine that these buildings as a group were largely conformist rather than innovative. The period of building in Britain leading up to the First World War can be regarded as a confluence of opportunity whereby industrial and colonial wealth provided high-quality materials to be worked by tradesmen belonging to a well-established and organised working class with extreme skill. Such conditions provided a context able to interpret and deliver standard designs using well-versed building practices with energy. The year 1919 is commonly understood to be the end date of traditional building practices in Britain. This reflects the lives, knowledge and relationships lost in the First World

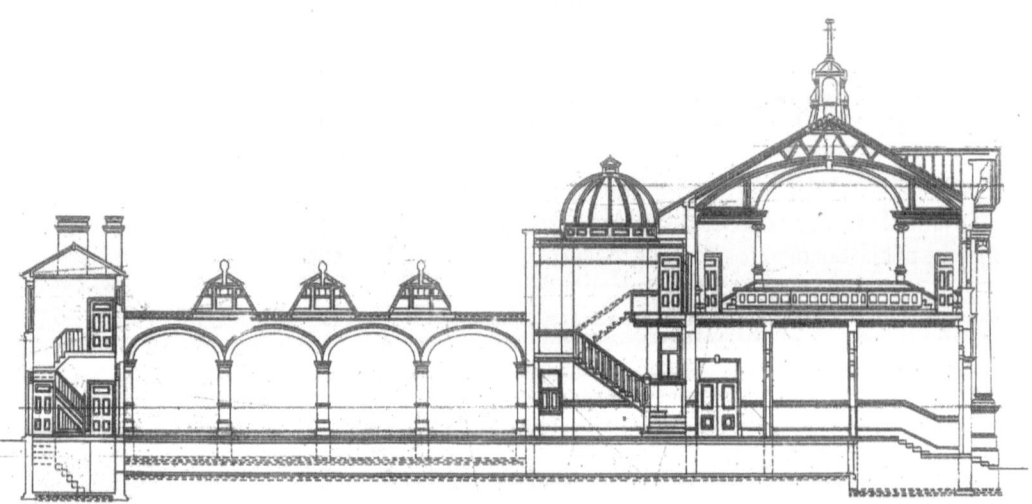

Figure 5.5 West Bromwich library, Stephen J Holliday 1907 Long section. Schofield, H. Borough Engineer and Surveyor, 1949.
(Courtesy Sandwell Metropolitan Borough Council Architect's Unit)

War and the abrupt re-focusing of imperatives thereafter. It is also, coincidentally, the year that Andrew Carnegie died.

Global Components

There are certain attributes of Carnegie library design that are almost universal. First, the relatively high cost of artificial light and the low cost of heat meant that library buildings were commonly designed to admit as much natural light as possible to benefit readers. To do so, they generally relied on the use of skylights. It was often the case in towns that a two-storey imposing front would conceal a single storey rear in order to maximise the admission of daylight from the brightest point in the sky. Figure 5.5 illustrates a typical top lit deep plan building at West Bromwich library. In describing a bid for Hammersmith library, Maurice B. Adams noted: "By this contrivance the cube, to save expense, was minimized, and the advantage of top lighting ... is obtained" (Ed., 1903b). The nature of a public library was famously described in J. W. Clark's (1894) Rede lecture as either "a workshop, or as a Museum". This characterisation is literally reflected by a description of Kettering library as having a "stack room lighted from the north on the usual weaver's shed plan" (Ed., 1903a) to admit light to its rear. In related work the research team have identified and collated standardised elements for incorporation in HBIM and quantified the use of rooflights and glazed domes evident in open Carnegie libraries in Britain today.

Liverpool's Toxteth (Southend Branch) is not a Carnegie funded library, however, it was opened by him in October 1902. This was the start (after Keighley) of his funding in England. Reportedly, he was so impressed by the £17,000 building as an exemplar of his preferred model of branch library (Ed., 1902c) as well as Liverpool's "virility" as a city to the free library movement that upon return to New York in a letter dated 16.12.1902 (Ed., 1902b) he pledged an almost unique unsolicited gift (Ed., 1905b) to the city for the complete cost of building their next planned £13,000 branch at Tuebrook (West Derby) and later a number of branch libraries. Indeed, as an architectural model, designed by city architect, Thomas Shelmerdine, it is paradigmatic of a twin gabled arrangement that subsequently predominates. Toxteth library today is also relevant to this study because it is located in an area with an Index of Mass Deprivation that is in the lowest decile in England. It is in good condition having been refurbished in 2008. Figure 5.6 shows views taken from a 3D pdf which can be downloaded (Boughanmi, 2020a) and navigated in full. It is generated from the REVIT™ model used to perform the LCA discussed later. Here, the isometric of the exterior and the separation of glazed areas are used to illustrate the extent of exposed surface areas of the building. This design became immediately typical, being originally designed to capture daylight effectively as a priority but now serving to reduce its thermal efficiency (and risk water ingress) today.

Figure 5.6. Isometric views from 3D PDF of Toxteth library showing (a) building envelope and (b) exterior glazed elements.
(Mahdi Boughanmi)

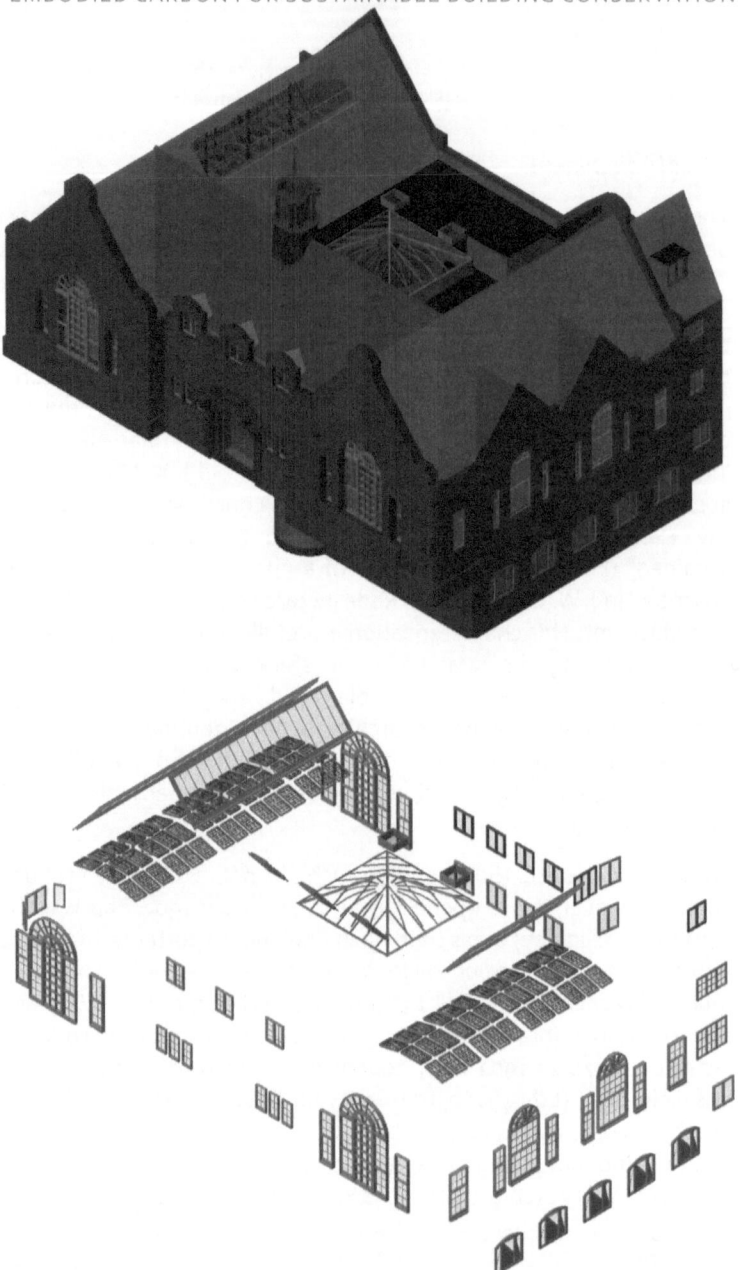

Local Materials

It is relevant to consider transportation distances with respect to current ambitions to acknowledge the embodied energy from existing buildings (Historic England, 2020). In visually reviewing the external wall materials used for British Carnegie libraries, it is clear that they were, in the main, locally sourced and regionally clustered. With few exceptions, the buildings are of solid masonry construction, predominantly built of either brick or stone. The materials chosen reflect their local availability and to an extent the geology of available building stones and clays of the ground. Evidence for this observation is backed up in detail through innumerable mentions in the architectural press. Lincoln library, for example, is one of the largest in England at 3044 m² (Ministry of Housing, 2020b) and was built of local Ancaster stone (Ed., 1917) whereas the Aberdeen central and branch libraries are clearly constructed of its local granite as those of Glasgow are predominantly of its red sandstone. Noting that, with few exceptions, architects tended also to be local to their buildings will account for some of this preference. They and their contractors are likely to have been familiar with the local materials and transportation costs would generally preclude the importation of heavy building material from distant sources. When considering the life-cost of these buildings as a group, this factor should be acknowledged more generally with respect to the quantity of embodied carbon that they represent. ˙

Altered Values

Changing Demands: Transformation of Socio-economic Contexts

The predominance of Carnegie libraries located in areas with low IMD today is not a coincidence but a direct reflection of a recognised crisis of post-industrial decline that divides society (Martin et al., 2016). The Scotch-American steel magnate favoured supporting the working man to help himself through the provision of public libraries in what were then industrialised areas. Today, particularly in England, the context of a service-dominated economy has determined that the majority of these library buildings now serve some of the country's most deprived communities. Our research has identified that in England, of a total of 147, 118 are located in areas with IMD deciles of 5 or lower. In Wales, of 14 open libraries remaining, 9 are in the WIMD decile of 5 or lower. Of those 19 buildings which have closed, moved or been re-purposed, 13 are in WIMD of 5 or lower. In Northern Ireland the 890 super output areas are ranked from 1 (most deprived) to 890 (least deprived). The three open Carnegie libraries are ranked within Northern Ireland's 2017 NIMD measures at 174 (Bangor), 55 (Falls Road), 54 (Lurgan). Notably, Old Park library, currently closed but previously volunteer run, sits in the second lowest area of all 890 in the whole of Northern Ireland. In Scotland, by contrast, the balance across deprivation areas is weighted towards the less deprived areas. This more even distribution is likely to reflect the higher proportion of buildings per capita that native Scotsman Carnegie funded there.

These data are crucial to include in this discussion because they underline the continuing, in fact enhanced, importance of the buildings and of their critical social role. It also highlights that the affordability of choice is likely to be limited and indeed that the pressures of sustainable management are significant.

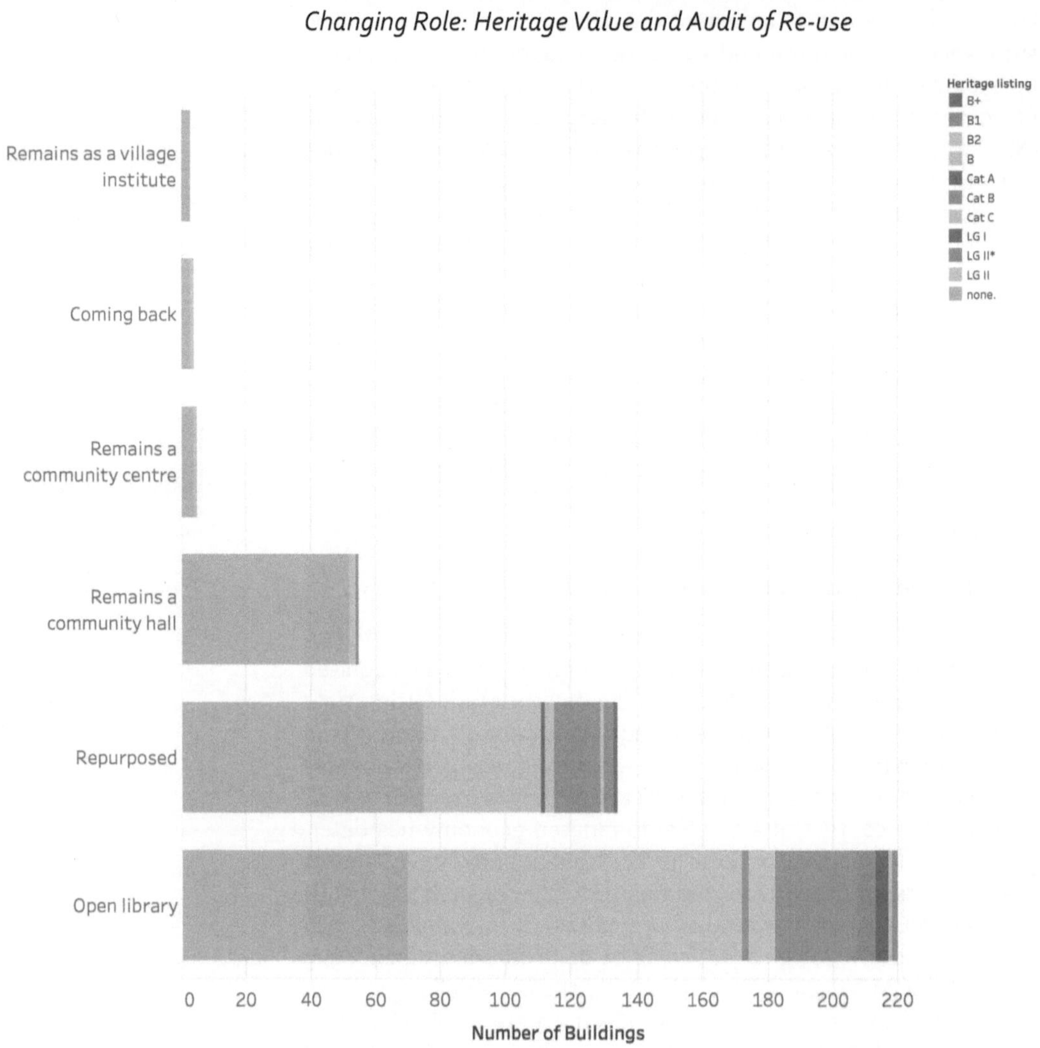

Changing Role: Heritage Value and Audit of Re-use

Figure 5.7 Bar chart quantifying functional status and heritage designation of buildings in the UK funded with Carnegie Library grants in 2020.
(Oriel Prizeman)

Since 1950 and accelerating from the 1980s onwards, the majority of the remaining buildings are now designated as heritage assets as plotted in Figure 5.7. This serves to both protect them from demolition but also at times to frustrate councils who perceive the upgrading of older buildings to meet new standards of wheelchair accessibility and energy performance as insurmountable. In addition to this, a willingness to adapt and move away from the designated purpose of the building may further enhance the prospects of its survival.

It is relevant to note the array of purposes for which these highly designated buildings have been re-deployed as illustrated in Figure 5.8. Although a substantial number of re-used library buildings have

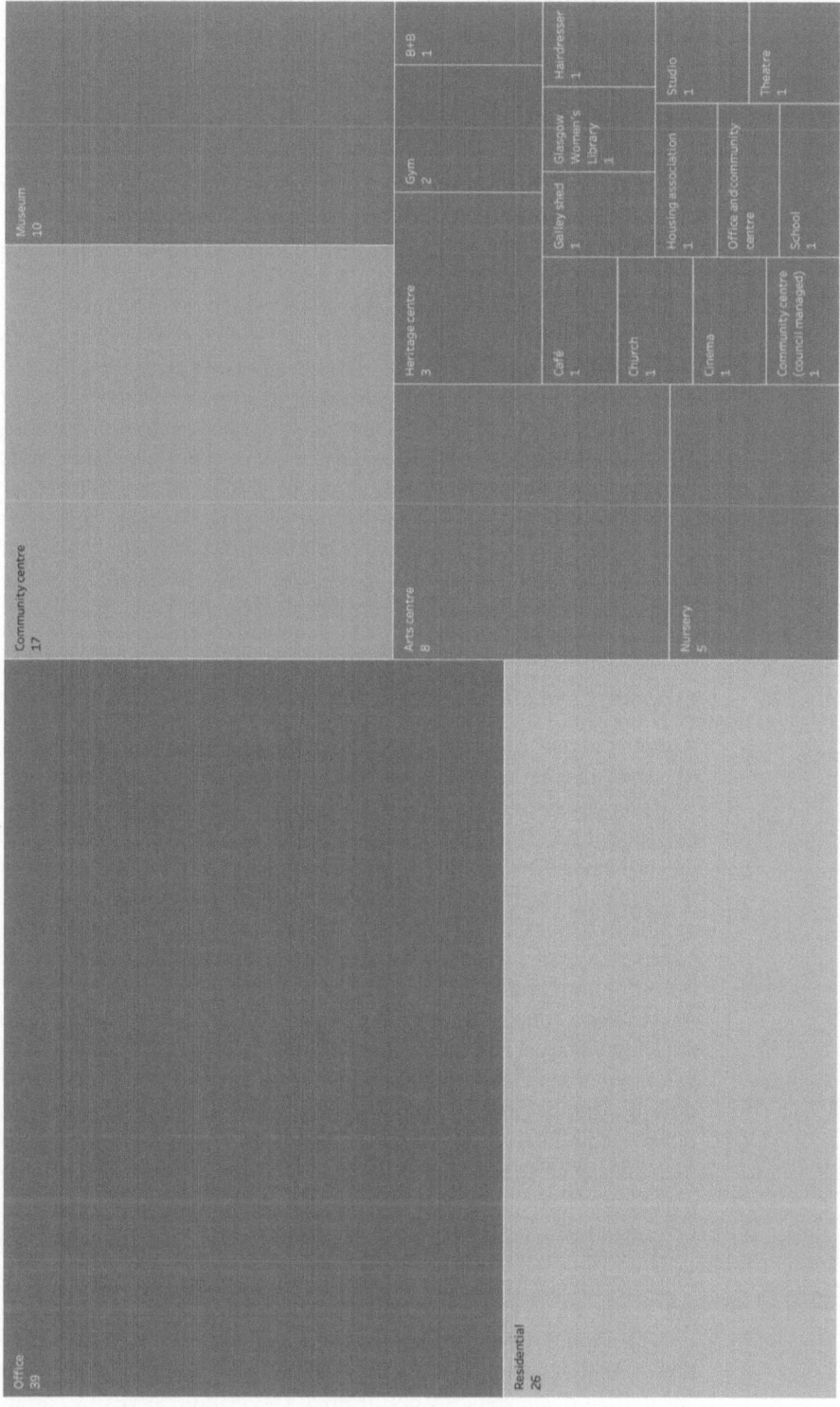

Figure 5.8 Treemap quantifying current use of former Carnegie library buildings that have been re-purposed. (Oriel Prizeman)

effectively been taken onto the private domain by being re-purposed as residential, and a large number have made use of the comfortable illumination and generous spaces of reading rooms to become offices, the majority of buildings have been re-used for purposes that can still be identified as serving the public interest. Although heritage officers might have opinions and planning departments may limit, for example the adoption of high street space for residential use, there is no regulation that can actively direct such outcomes, in effect, the re-use of a building is open to the developer or owner's imperative much more than that of the planning authority. Such movements are subject to much wider levers, predominantly these are economic, so the vulnerability of once publicly accessible library buildings to be replaced with something much less amenable to the communities they once served is high. Nevertheless, the table of re-use does show a tendency in the case of the Carnegie library buildings in Britain that have been re-purposed, to be understood as socially beneficial environments. It is clear that they have generally been re-deployed for uses that might be associated with the role of a public library in its wider sense, as a place that served the community. Thus, there are numerous educational uses such as schools and nurseries, but also a number of rehabilitation related community centres. The identification of the qualitative environmental factors that stimulate such a coherent response is a subject for further research.

Changing Estimations of Performance: Measures of Efficiency

As noted above, there are assumptions that heritage assets are liabilities with respect to energy use as their adaptation can be complex, however the evidence collated here suggests this is not the case, indicating that political will or bias may still have a greater impact. Using newly accessible data for all Display Energy Certificates in England and Wales (Ministry of Housing Communities & Local Government, 2020), it is possible to compare the measured operational rating bands of 107 of the open Carnegie library buildings with DECs that have been logged to date in Northern Ireland, England and Wales. Asset ratings on Energy Performance Certificates (relevant here to libraries which are rented spaces and those in Scotland) are calculated values whereas Operational Ratings on DECs are based on metered data benchmarked to a target for the building type. They are not directly comparable yet are both designed to deliver unambiguous measures of performance to stakeholders and managers. There is leverage in evidencing poor energy performance in support of either closing or moving a library service.

The benchmark given to all is "Cultural Activities". The typical building in the UK is expected to be rated between D and E (Davies & Chartered Institution of Building Services, 2009). Figure 5.9 shows there are six libraries here which achieve an Operational Rating Band "B" in their display energy certificates. This is impressive in so far as all six are listed Grade II, all are at least 110 years old, none have been significantly modified and all are predominantly naturally lit. They cover a range of attributes: Of these only Malvern (1906) is a fully detached building. Ilkley (1907) is attached to the town's public offices and assembly hall (News., 1905). Kayll Road in Sunderland (1909) is community managed and although it does have a suspended ceiling it retains a large rooflight, Folkestone (1910) is an extension to an existing building and Leicester central (1905) is also attached to council offices. Northampton central library (1910) is one of only four of all the Carnegie libraries to be terraced. At the other end of the scale, only 3 libraries received a G rating, Harlesden, Pontypool and Brentford and only four are F rated but have no data so this would appear to be a default setting.

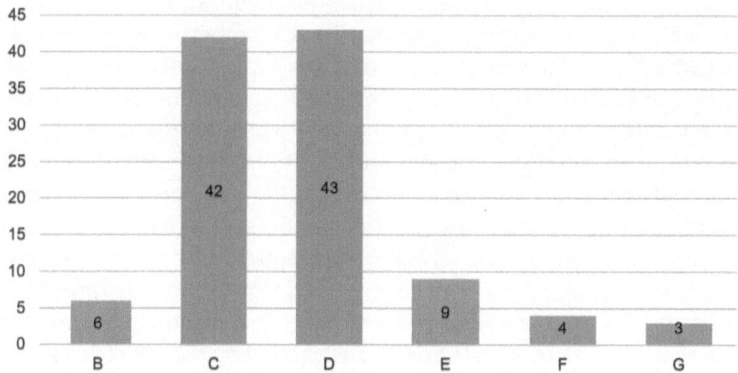

Figure 5.9 Chart plotting Operational Rating of all open Carnegie libraries England Wales and Northern Ireland with DECs 2020.
(Oriel Prizeman)

In terms of heating fuel, Leicester central is the only building here to have district heating, Abergavenny and Meadows library in Nottingham are the only two to list renewable sources (solar and solar PV). Twenty-three have air conditioning, only eight list Heating and Mechanical Ventilation although it is known that the majority were designed to incorporate it. Ninety-five are all listed as Heating and Natural Ventilation, one as "Mixed-mode with mechanical ventilation" and three Mixed-mode with natural ventilation. Floor areas range from 109–5000m².

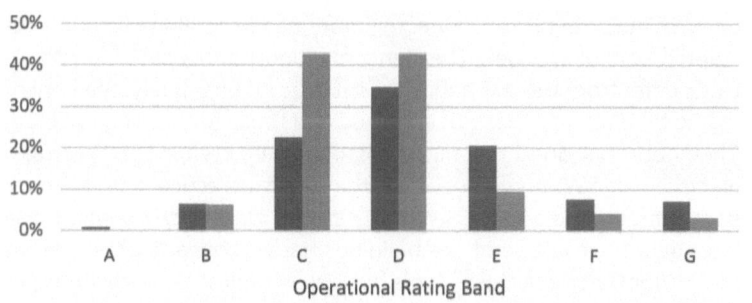

Figure 5.10 Chart plotting all public buildings with DECs in England and Wales Q1 2008–Q1 2020 and all DECs for Carnegie libraries
(Oriel Prizeman)

Figure 5.10 demonstrates that 84% of the Carnegie library buildings here are deemed to be D or above, whereas the score for all public buildings with DECs in England and Wales rated above band D is only 65%. In contrast with DECs for all public buildings in England and Wales Q1 2008–Q1 2020 (Ministry of Housing Communities & Local Government, 2020), the profile of the Carnegie libraries is not radically worse as might be anticipated.

The Operational Rating can be expressed as:

$$OR = 100 \times \frac{Building\ CO_2\ emissions\ /\ building\ area}{Typical\ CO_2\ emissions\ per\ unit\ area}$$

However, in our sample the operational bands are not correlated to the same typical usage benchmarks, although all but one (under "office"), cite the same overall benchmark property type; "cultural activities". The research observed that the "typical" measures of fossil fuels used as benchmarks range from 138–257 kWh/m² per year with a few extreme exceptions (39 and 307) which are assumed to be errors. This correlates closely with latitude, however, measured thermal energy use did not. This observation is important as it highlights the risk of presumption in benchmarks. Whereas latitude can be seen to correspond closely with the gradual adjustment to benchmark data applied for thermal energy use, Figure 5.11 illustrates that it has no discernible influence on either actual electrical or thermal usage across the range of varied sizes of building. For electricity, the "typical" measures on the DECs for Carnegie libraries ranged from 47–124 kWh/m² per year.

CIBSE Guide F benchmarks set in 2012 for existing public library buildings suggest that good practice is fossil fuels 113 kWh/m² per year and electricity 32 kWh/m² per year but that typical practice is 210 kWh/m² per year fossil fuels and 46 kWh/m² per year electricity (Ed. Butcher, 2012). The inadequacy of CIBSE building type benchmarks was highlighted in Hong's PhD research (2015). Subsequently, tools being developed in collaboration with UCL and available from CIBSE (2020) have created a dynamic energy benchmarking tool dashboard, currently in BETA, which to date includes data for 275 existing public library buildings nationally based on DECs. We might presume that our 107 DECs found for Carnegie library buildings alone form a substantial part of this set. The benchmarking figures based on these DECs differ from the last published set in 2012: Effectively, the benchmarks for electricity have been raised and those for heating lowered. They note that good practice is fossil fuels 85 kWh/m² per year and electricity 54 kWh/m² per year but that typical practice is 117 kWh/m² per year fossil fuels and 76 kWh/m² per year electricity. Against these later data the average of the buildings here is above the "typical" at 160 kWh/m² per year fossil fuels but below typical at 62 kWh/m² per year for electricity as illustrated in Figure 5.12.

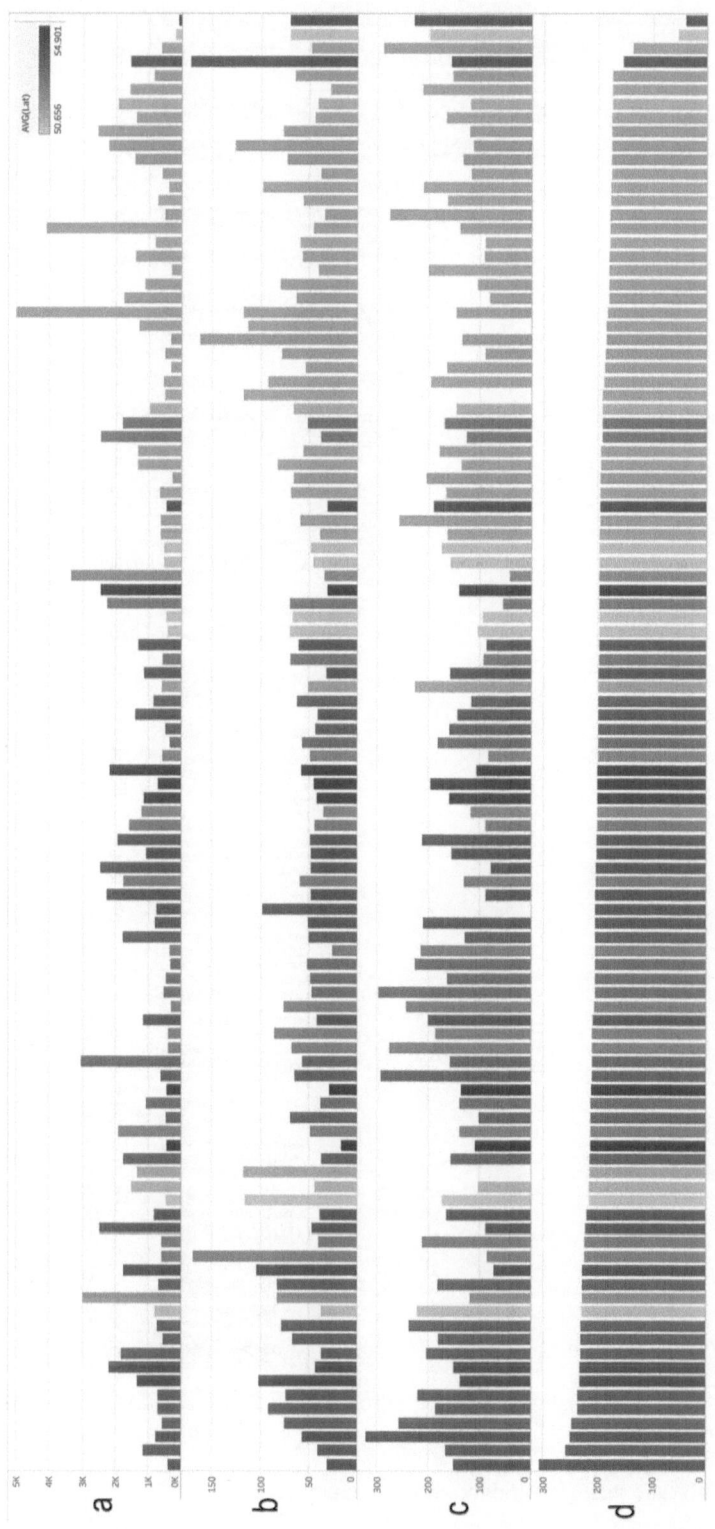

Figure 5.11 Chart shaded to indicate latitude (darker is further North) against data for all open Carnegie library buildings with DECs in England, Wales and Northern Ireland aligning (a) Total Floor Area m² (b) Annual Electrical Usage kWh/m² per year (c) Annual Thermal Fuel Usage kWh/m² per year (d) Typical Thermal Fuel Usage kWh/m² per year benchmarks used in each DEC. (Oriel Prizeman)

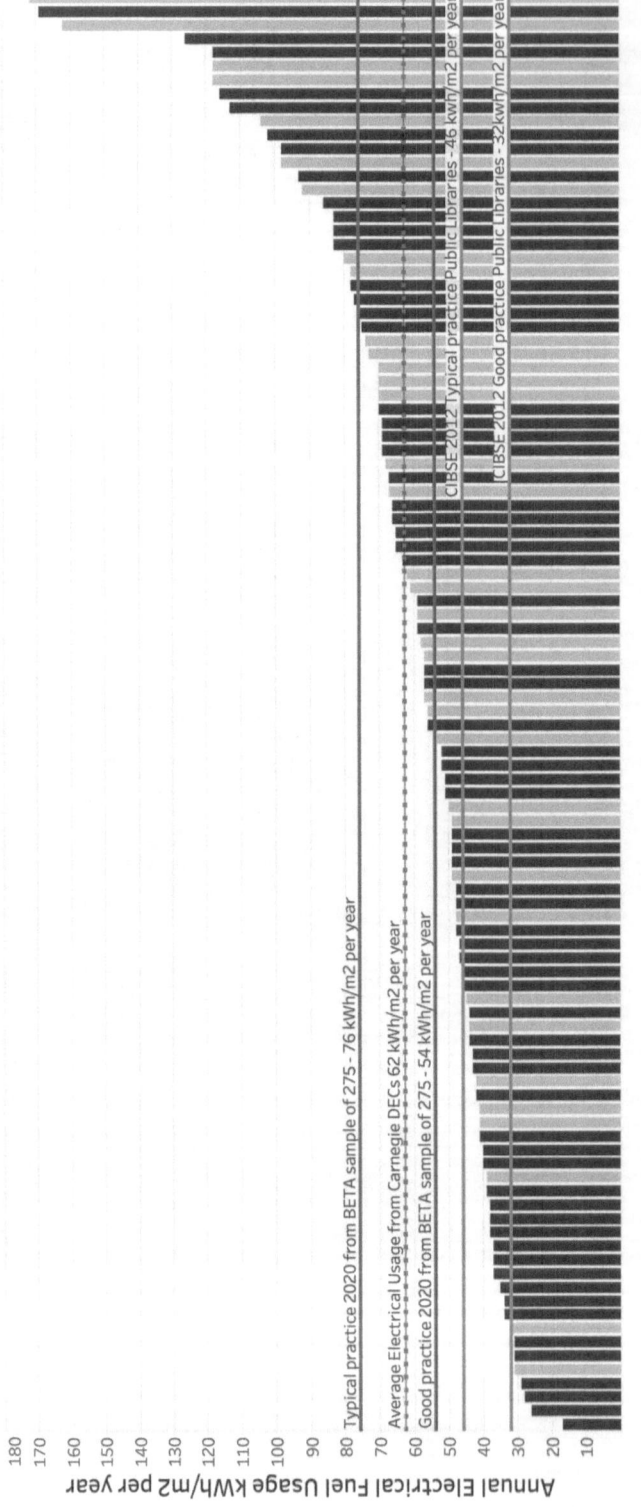

Figure 5.12 Chart plotting annual electrical fuel usage for open Carnegie libraries with DECs against 2012 and emerging 2020 CIBSE benchmarks for public libraries. Presence of skylights indicated by darker bars.
(Oriel Prizeman)

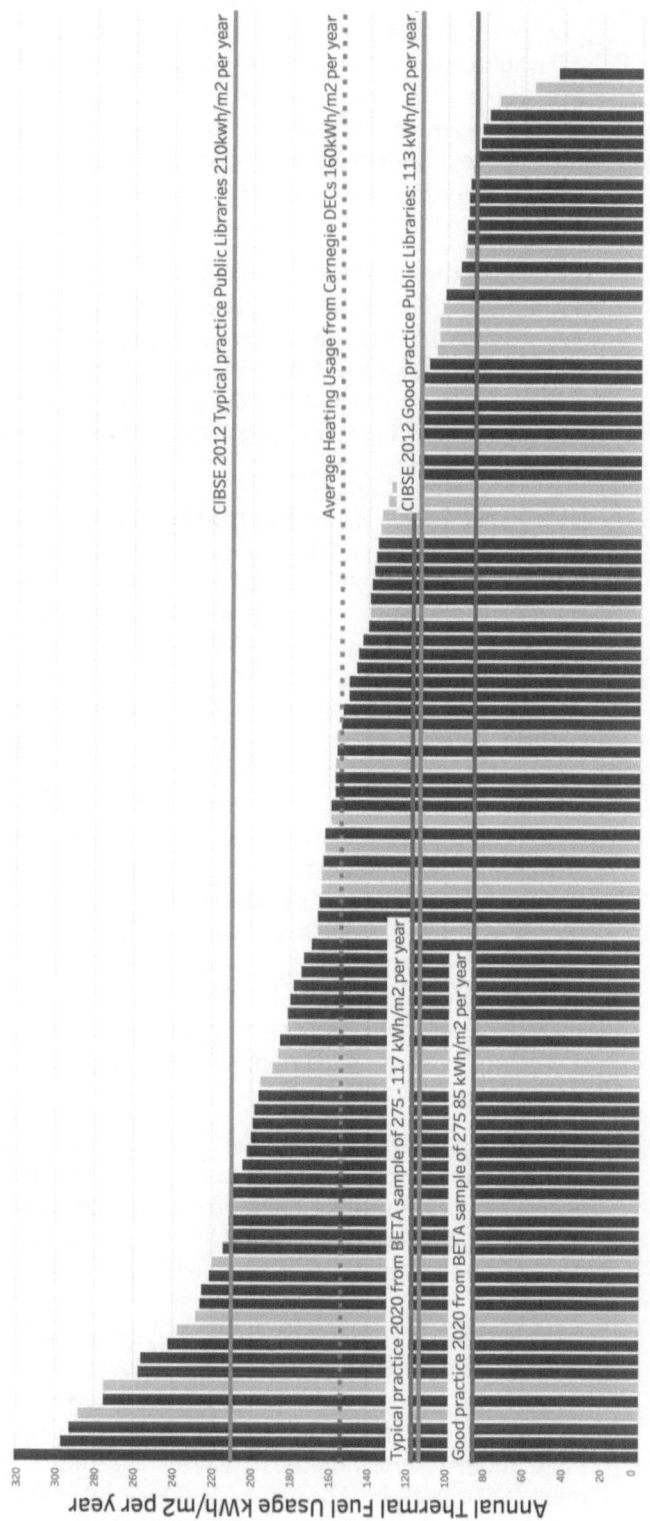

Figure 5.13 Chart plotting annual thermal fuel usage for open Carnegie libraries with DECs against 2012 and emerging 2020 CIBSE benchmarks for public libraries. Presence of skylights indicated by darker bars.
(Oriel Prizeman)

Despite the number of visible lights switched on in daylight hours, it is possible to speculate that the predominance of relatively high levels of daylighting contributes to the relatively reasonable use of electricity in the Carnegie library dataset. Forty-four of the 107 libraries that are below average electricity use have skylights, whereas only 25 of those above do as indicated in Figure 5.13. In addition, several others below average have either highly glazed walls, glazed domes or clerestory lights.

For heating the correlation is not as distinct, 36 below average have skylights, whereas 32 above average have them – see Figure 5.13. This may indicate that skylights are not as great contributors to heat loss as presumed because of the solar heat gain they enable. The higher thermal figures from the DECs of open Carnegie libraries here indicate the margin of improvement required to match the most recent standards of good practice.

Although EPC Asset Rating Bands and DEC Operational Rating bands are not directly comparable, the graphic impact of the colour-coded range, A–G, is illustrated in such a way as to give an unambiguous impression of best to worst that is potentially persuasive to decision makers. When looking for modelled estimations for the energy performance of these buildings, there are obviously fewer available since the DECs are the standard requirement for public buildings in England, Wales and Northern Ireland. However, Energy Performance Certificates are mandatory instead of DECs for public buildings in Scotland. Nevertheless, using postcode lookups only 12 could be found of the 56 open Carnegie libraries there. A further 14 are also available for Carnegie libraries in England and Wales. Using this dataset of only 26 buildings it is evident in Figure 5.14 that the calculated energy use of these buildings appears to be higher on the scale than the measured energy use determined in the DECs. The data include seven buildings for which no Primary Energy Value has been entered, indicating that these are presumptive figures. It is assumed that the one library that is A rated (Wombwell) is an error as it has no obvious significant modifications.

Nevertheless, overall, the data illustrate that for existing buildings, the EPC method of modelling tends to indicate relatively higher levels of energy and CO_2 emissions than the measured data provided through DECs. This is concerning precisely because the opposite has been demonstrated to be true for new buildings that are being designed based on models that turn out to be over optimistic relative to their measured data of consumption. The comparison indicates that there is a bias evident in modelling tools which favour projected performance of new building products and workmanship over existing materials and workmanship for which there is a significant potential to generate more accurate observations. They also indicate that frugal management can have as great an impact as significant alterations to a building's fabric or services.

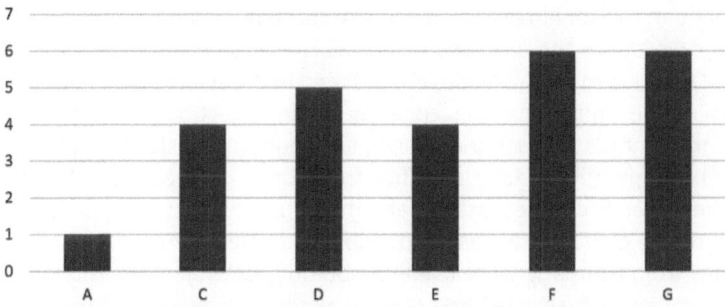

Figure 5.14 Chart plotting numbers of open Carnegie library buildings by 2020 in Scotland, England and Wales with an Energy Performance Certificate within Asset Rating Bands.
(Oriel Prizeman)

Drawing Life Cycle Indicators

As noted above, attempts to calculate the life cycle of an existing building completely are not feasible, owing to a lack of data and also not helpful in determining future expenditure, for the purposes of projection, it is conventional to strike off previous expenditure as "sunk costs" (Flanagan, 1989). However, it is still relevant to seek to derive indicators from a reference model for a set of buildings which share so many attributes. Using Toxteth library, which was identified by Carnegie as an exemplar, we are able to provide some signposts.

LCA: Toxteth Library – Materials and Components

Figure 5.15 Toxteth library: Cross-section of laser scan.
(Camilla Pezzica and Giovanni Bruschi)

A parametric model of the existing Toxteth library derived from a laser scan Figure 5.15 was built in REVIT™ and using the OneClick LCA™ plug-in, the take-off of elements was imported in an attempt to create a life cycle assessment projected for 60 years in accordance with EN-15978. Assuming that historic means of transportation, by boat, train or horse-drawn barge or carriage were beyond the scope of calculation, the aim was to model the building as if constructed today, using materials as closely matched to the actual building as possible. The immediate challenge lay in the selection of appropriate equivalent materials. For example, although the substantive quantity of brick might be readily identified from the UK, there is no option for a lime mortar or lime plaster that is manufactured in the UK, only from Germany or France. It was not possible to locate natural slates, so roof tiles from Germany were used. In addition, of course, the quantities and types of fuels used and the processes of manufacture for steel, glass, kiln drying, lime-slaking, timber seasoning, slate mining, stone quarrying or tile-firing are all more or less incomparable in terms of their energy or carbon costs. Added to this the chemical composition of finishes; paints, varnishes, polishes and stains is generally completely altered. The outcome was therefore fairly unsatisfactory as it was so reliant on substitutions, some of which are quite eccentric owing to the changed standards of building technology today.

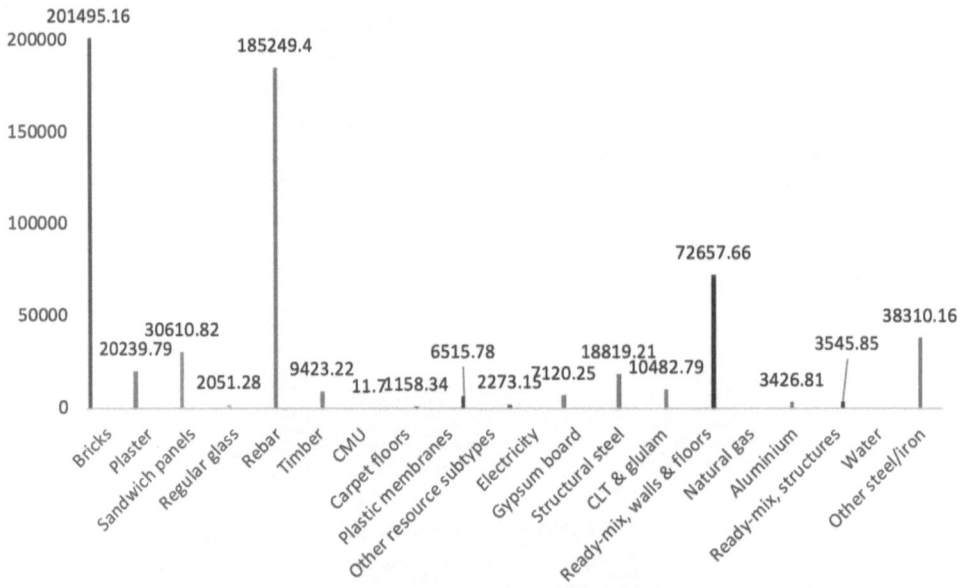

Figure 5.16 Toxteth library: Total life cycle impact by resource type and subtype. Global warming KgCO$_2$e excluding energy and electricity use.
(Oriel Prizeman)

There are, however, some critical details to highlight. The analysis, which can be accessed here (Boughanmi, 2020b), obviously showed 60 years of energy use to be the greatest contributor to future projected emissions. However, excluding that data, Figure 5.16 above sets out the embodied carbon in attempting to construct a similar building today, albeit including some unusual choices as noted above, there are some relevant lessons in principle. Excluding Rebar (steel reinforcing bar in concrete is not a known constituent in any event just a presumption from the modelling software REVIT™), the quantity of embodied energy attributed the large volume of masonry stands out. Since both the form and the method of construction are typical to a significant number of buildings, it is worth reflecting on the impacts of this.

Suggested Adjustments

External Walls

The record of the building's opening in the architectural press (Ed., 1902a) determines its key materials – this was a common practice in reporting so such data are frequently accessible. We know that its external walls were constructed of Ruabon bricks with Cefn stone dressings, these are adjacent sites near Wrexham in North Wales, 58 miles south of Toxteth. The slates on its roof were Green Cumberland slates, presumably from Borrowdale, 102 miles north of the site. Furthermore, looking at the map, it is likely that the weight of these materials on these journeys was likely to have been at least partly waterborne. These short distances from cradle-to-gate are accounted for here only in part, however their contribution to the weight of materials used is quantified – with external walls accounting for 39.7% of the overall mass of the building. It is not unreasonable, regarding the map of exterior materials used for all Carnegie libraries in Britain and acknowledging the similarity of their construction methods, to extrapolate that the distance of building materials from the sites was generally kept to a minimum and that the use of local stone, which was frequently praised, if nothing else because it benefitted the livelihoods of local men.

Figure 5.17 Toxteth library: Parametric model detail and plan detail showing external wall construction.
(Mahdi Boughanmi)

The greatest volume of material used is for the wall construction. These measure 535 mm thick and are deemed to be built of two 225 mm skins of solid-bonded brickwork finished internally with lime mortar and plaster (Figure 5.17). Providing a small cavity between the two skins was an emergent practice used to prevent direct ingress of moisture, "hollow walls" are described in contemporary specification guidance (Macey, 1898). Based on the BR 443 the U-value of each 225 mm skin would be 1.52 whereas Rye and co have monitored a similar construction (no 20a) and found a figure of 1.48 p. 36 (Rye, 2012). As the difference is not huge, BR 443 U-value of 1.52 which is based on BuildDesk™ calculations was used. The modelled U-value of the whole wall assembly in REVIT™ is 0.3448 W/(m²/K). The walls' thermal mass is calculated as 57.72 kJ/K and is likely to benefit the thermal comfort of the building year-round. Upgrading the thermal efficiency of the walls further is generally not the optimal solution; external insulation would be ruled out for aesthetic reasons relating to listing status. Cavity insulation is no longer deemed advisable, internal dry lining might in some cases be possible without damaging the historic fabric, however, the benefit of thermal mass would also be lost.

It should be stated that the standard use of this quantity and quality of material whilst technically possible, was an extravagance not repeated after the economic contraction post–First World War. It would never be specified today simply on grounds of cost and the use of floor space given over to non-profitable structure. The durability of such practices of solid masonry construction is evident in the physical survival of every building. In order to meet modern standards, such walls would only be faced in a single skin of brick for aesthetic reasons, otherwise, they would be composed of a frame only substantial enough to meet structural requirements and insulated to meet the modelled energy performance targets of today together with an anticipated service life of perhaps 60 years. Taking into account the additional service life of the stone dressings and sills around the window openings that have also endured, it is hard to anticipate that aluminium, concrete or timber alternatives that would be chosen today would not require significantly more frequent maintenance and replacement.

Floors

Floor finishes that require frequent replacement have significant impacts on life cycle assessments. With the exception of the first-floor gallery and carrels, the interior floors at Toxteth are of solid wood block parquet, a typical solution of the era (Figure 5.18). The floors require polishing and have evidently been sanded and re-finished in the 2008 refurbishment, however, other than general cleaning, they have survived for 118 years. The quality and colour of their finish is also a recognisable visual asset. As with the wall material, today, such solid construction is unaffordable and could not be justified for a building of this type except if specified in a laminated and therefore more

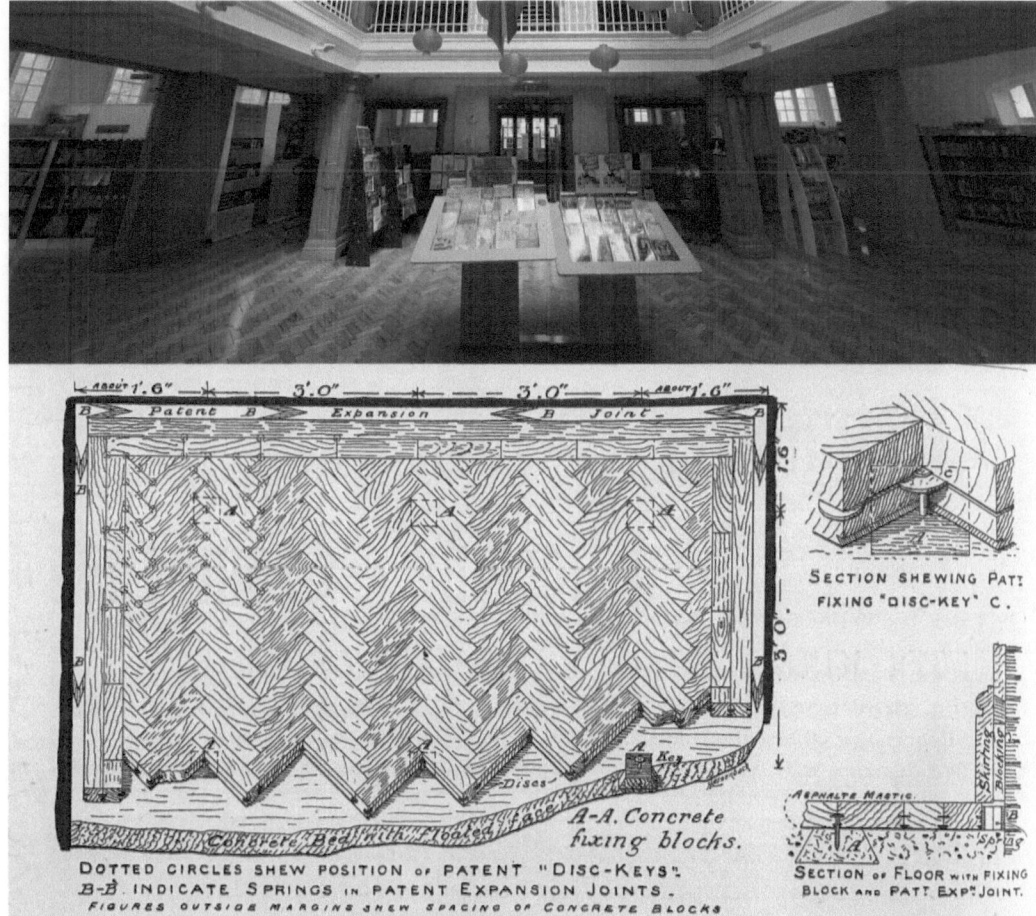

Figure 5.18 Toxteth library: Interior showing parquet floor (Oriel Prizeman) and detail of typical construction.
(Sears, 1893)

vulnerable form. Requirements for acoustic attenuation in libraries would generally dictate that a softer floor finish were specified, perhaps a carpet, a vinyl or linoleum floor. These would have the same cleaning and maintenance demands as the woodblock floors but an anticipated service life of 5–10 years.

Rooflights

Toxteth library has a footprint of 620 m², its roof slopes are typical in including 124 m² of rooflights, see Figure 5.19. It is understood that all glazing was draught-stripped and updated in 2008. The critical vulnerability of the top lights as being the source of valued daylight but the most direct exit of heat is obvious. To this end the replacement of the external components of the rooflights with triple-glazed sealed units

Figure 5.19 Toxteth library: Laser scan cut away showing internal and external glazed layers in top-lights.
(Camilla Pezzica and Giovanni Bruschi)

has been considered. It should be noted that as at Toxteth, the larger rooflights and glazed domes were not simply a single construction in the line of the roof, they commonly had a lower layer internally, often with decorative stained and obscured glass leaded lights. These served to diffuse the unwanted over-heating of direct sunlight. Having observed librarians in the North of England sheltering beneath a plastic gazebo that they had erected to shade themselves from the unprecedented heat of summer 2018 inside their refurbished library, it is important for managers and architects to note that the obscured glass layer that had been replaced with a modern clear rooflight did not simply serve an unnecessary decorative purpose.

Services

By far the greatest contributor to the CO_2 emissions of Toxteth library modelled over the next 60 years is its projected energy use. Fossil fuel heating is calculated at 31,272t or 67% and electricity use at 929t or 20% of the total CO_2. Swapping the source of heat for the most efficient available option in the model (biodiesel, vegetable oil based, SAP) reduces its contribution to 268t or 15%. Were the source of electricity to be decarbonised by, for example, connection to the world's largest offshore windfarm, at Burbo Bank adjacent to Liverpool, these figures could be amended further with little or no physical impact on the library building. Whilst this is not a universally available opportunity, it is illustrative of the increasing potential to set the problem in context.

Discussion – Balancing Goals and Generalisable Implications

Balancing Goals and Relative Measures

Data collated here relates to quantified economic, social and political conditions but also to related implicit qualitative measures of cultural heritage values, environmental wellbeing and regional socio-economic inequality. The force by which these imperatives are addressed through political policy and ultimately fiscal decisions is often distanced by levers informed by unilateral goals. Nevertheless, a significant number of community projects demonstrate that there are also bottom-up forces at work which are likely to innovate through improvisation. Observing the distinction between global and local drivers for change is thus utterly critical when applied to a set of buildings despite their apparently being erected with such uniform aspirations.

With such small numbers of top-ranking libraries (Figure 5.9), it is possible to consider each case in turn. Sunderland's Kayll Road library faces specific challenges, the library service was closed by the council and the building has been given to the local community, it is now one of a growing number to be run by an entirely voluntary group. In this area, where the median income is in the 4th decile and employment in the 3rd (Ministry of Housing, 2019), it is a reflection of the perceived value of the facility that its six computers available were provided by local contributions. The probable cause of Kayll Road's inclusion in the highest ranking is the frugal nature of its management. Malvern, by contrast rests in the 9th decile for deprivation, its inclusion in the least resource consuming band is probably an error based on the computation of its floor area which stands out as one of the largest Carnegie libraries in Britain on the DEC at 5000 m², from Ordnance Survey it would appear more likely to measure around 1100 m². It is likely that the unusual embedding of the building (together with district heating at Leicester) account for the better reported running costs in Folkestone, Leicester and Northampton central libraries. Ilkley is a more typical building and its high score perhaps reflects a degree of frugal management. In common with many buildings it benefits from "strong top lighting" (News, 1905) and was typical in being designed with mechanical ventilation and vaulted ceilings.

The libraries at the bottom of the performance scale are not discernibly different in their physical characteristics from the main. However, it is noticeable that a disproportionate six of the seven in categories F and G are in Greater London: Lea Bridge, Harlesden, Walthamstow, Brentford, Willesden and Teddington. Given that the buildings are not particularly different physically from others in the UK, an observation may be either that being frugal with heating is at a lower priority than in other parts of the country or that as with Walthamstow's example in the introduction, they are simply very intensively used. Teddington is in the least deprived area of any open Carnegie library and is one of only two to be in the top decile for IMD in England. It is a fairly typical double gabled design and is only different from its peers in terms of its internal decorative plasterwork which is unlikely to significantly impact its energy use. Notably two of them (Harlesden and Walthamstow) were not originally Carnegie library buildings, they were built earlier and had extensions funded by Carnegie. The original part of Willesden library is dwarfed by an enormous extension so the data relating to it have little bearing on the old building.

These relatively small numbers of extremes are exceptions, rather it is the consistency of the cluster of median values around C and D ratings that are important to acknowledge. These measured data of actual operational energy consumption are impressive for a building stock of this age, especially with respect to the challenges outlined above that are inherent in the original aspirations of their designs. Bearing in mind that these data are comparative with all other buildings of all ages of this type, they would appear to reflect efficient management practices. It may also suggest that the almost universal preponderance of high levels of natural lighting within Carnegie library buildings is still a benefit to operational demands even though the energy consumption of artificial lighting appliances has been so reduced in recent years.

Generalisation

The findings regarding the relationship between measured energy use and current as well as emerging benchmarks confirm robustly that this cohort of buildings are not performing as badly as presumed for their age, particularly for the use of electricity. Reducing the consumption of energy for heating from 160 kWh per year (the current average of this cohort of 107 Carnegie library buildings) to that 85 kWh/m²K per year attributed to the emerging "Good practice" benchmark from the CIBSE's project in BETA, is more challenging.

However, six buildings in our dataset with attributes that are shared and of very different scales, appear to achieve this already, all are fuelled by Natural Gas. They are Kings Heath, Rushden, Tower Hamlets Local History Library & Archives (formerly Mile End library), Burnley, Harrogate and Walsall. All are Grade II listed and Figure 5.20 illustrates their varied scale and appearance. Of particular note is the exemplary recent refurbishment of Walsall library, which has achieved annual energy consumption for heat of just 55 kWh/m²K. That this is possible for one, should, by virtue of all the observations made during this research of common features, make it feasible for others to follow.

Table 5.1 indicates the current predominance of Natural Gas as a heat source for these buildings. It could be presumed that external factors will influence this in coming years. As noted initially, there is potential for increased reliance on lower tariff forms of energy to benefit their operational usage and increasing reliance on renewables which can

Figure 5.20 Existing Carnegie library buildings with DECs exceeding emerging CIBSE 2020 "Good practice" benchmark for heat energy use of 85 kWh/m²K per year in public libraries (a) Kings Heath – Arthur Gilbey Latham, 1906 (b) Rushden – William Beresford Madin (Town Surveyor), 1905 (c) Mile End (James Knight originally 1862, Extension by MW Jameson, Borough Surveyor funded by Carnegie), 1906 (d) Burnley – George Hartley and Arthur Race (Borough Engineer), 1930 – Carnegie UK Trust (e) Harrogate – Henry Thomas Hare, 1906 and (f) Walsall – James Glen Sivewright Gibson, 1906.
(Oriel Prizeman – all images)

Table 5.1 Heat sources

Natural Gas	Oil	Grid supplied Electricity	District Heating
98	3	5	1

(Oriel Prizeman)

be delivered at district level is not detrimental to the visual setting of the building. As the photograph of Walthamstow at the beginning illustrates, many libraries still operate artificial lights even though their levels of natural illumination are more than adequate. The comprehensive installation of sensors to time artificial lighting would be a low-cost adjustment to make. The iSERVcmb project has demonstrated through monitoring how calibrating the timing of heating controls in buildings with significant thermal mass can initiate energy savings of up to 43% (Knight).

Another potential is to re-design the mechanical and ventilation systems taking note of their original intentions. As Table 5.2 indicates, 90 of these 107 buildings are listed on their DECs as being naturally ventilated.

Table 5.2 Building environment

Air Conditioning	Heating and Mechanical Ventilation	Heating and Natural Ventilation	Mixed mode with Mechanical Ventilation	Natural Ventilation Only
4	7	90	4	1

(Oriel Prizeman)

However, our survey identified ventilation turrets or towers associated with mechanical ventilation systems on 90 of 224 open library buildings, including Toxteth, a further 71 have external vent grilles visible which may indicate that they also had the same system but that the turrets have been removed (Prizeman et al., 2020). They were typically designed to use mechanical ventilation with fans in turrets over vaulted ceilings with inlets drawing warm air through the wall behind the radiators below as modelled in principle in recent work (Prizeman et al., 2020). The principle is neatly described for Montrose library: "the building is heated by low-pressure hot water, and pure air is secured by two electrical fans – one above the lending library, the other in the flèche above the recreation-room, which effect a complete change of air every fifteen minutes without draught" (Ed., 1905a). Although drawing in such quantities of fresh air is counter to common ambitions of air tightness, it is possible that modern heat recovery systems could be devised to systematically make use of these existing features.

In seeking to meet the lowest of the various benchmarks for electricity consumption, the CIBSE 2012 Good Practice level of 32 kWh/m²K per year from the current average of 62 kWh/m²K per year, there are more exemplars in our sample already below this level. These are shown in Figure 5.21; the previously mentioned Kayll Road together with Stirchley, Annfield Plain, Herne Hill, Heckmondwike, Ilkley, Middlesbrough and Batley. All are Grade II listed except Heckmondwike, which has no heritage designation. They are variously council and community managed and cover a wide range of economic contexts. Again, the potential for generalisation is self-evident.

Conclusions

Modelling and measuring energy use here identify a reverse principle of normal assumptions of performance. The metered data reveals consistent thrift in management and durable economic operation learned over time. Identifying the use of local materials in these proto-modern early standardized buildings is a significant finding. Mapping the local and the global paradigm demonstrates some irony in that the specific was generic (American furniture fittings and library plans) and the generic was specific (the use of local stone).

Figure 5.21 Existing Carnegie library buildings with DECs exceeding emerging CIBSE 2020 "Good practice" benchmark for electrical energy use of 32 kWh/m²K per year in public libraries (a) Kayll Road – Hugh Taylor Decimus Hedley, 1909 (b) Stirchley – John P. Osborne, 1908 (c) Annfield Plain – Edward Cratney 1908 (d) Herne Hill – H. Wakeford & Sons, 1906 (e) Heckmondwike – Henry Stead, 1911 (f) Ilkley – William Bakewell, 1907 (g) Middlesbrough – Sir Thomas Edwin Cooper, 1912 (h) Batley – Walter Hanstock & Son, 1907.
(Oriel Prizeman – all images)

Whereas Hong (2015) noted that establishing benchmarks for the energy use of a typology is generally difficult, with the large quantity and limited era of building during which the Carnegie library pro-gramme was delivered in Britain, it is possible to make steps towards offering informed statements in principle since the buildings demon-strably deploy a range of standardised components and features. There is potential for these methods, which have been based on his-torical research and computer vision tools, to be expanded to other sets of buildings or specific technological practices. The importance of considering ambitions for lowering carbon and energy is undeniable, however, reasoning through the imperatives of original design inten-tions is paramount to better decision making.

For these buildings, their potential social value, current heritage listing status and universal conditions of operating under tight financial constraints are also shared circumstances in addition to requirements to maintain accessibility and meet lower emissions targets. These mutual conditions can readily be assumed to have similar impacts on future attempts to reduce energy and carbon costs. Here, using a single but exemplary building as an archetype, key components and opportunities have been identified whilst common traits with respect to the use of materials are mapped to provide evidence in support of generalizing these findings with confidence.

The findings suggest that a cautioned approach to decision making should be adopted which seeks to make use of existing potentialities:

- Future research to develop generic ventilation and heat recovery strategies relating to existing equipment and built forms could be developed
- to prioritise the replacement of fossil fuel sources of energy with low carbon or renewable alternatives as they become available
- to consider the installation of additional insulated glazing over rooflights

and not to:

- Drastically alter the daylighting design of these buildings or
- consider that they do not compare favourably with newer buildings in terms of energy performance.

Where it is necessary to move the library service on, the findings demonstrate that future uses are most successful that are aligned to celebrate the particular environmental qualities of publicly accessible environmental conditions offered. Fundamentally, this paper calls for decision makers not to discount the durable quality of materials and workmanship and the ongoing contribution to wellbeing of these civic buildings as a given.

Reference List

Arpke, A., & Hutzler, N. (2005). Operational life-cycle assessment and life-cycle cost analysis for water use in multioccupant buildings. *Journal of Architectural Engineering*, *11*(3), 99–109. https://doi.org/10.1061/(ASCE)1076-0431(2005)11:3(99).

Avrami, E. (2016). Making historic preservation sustainable. *Journal of the American Planning Association*, *82*(2), 104–112. https://doi.org/10.1080/01944363.2015.1126196.

Battle, G. (2010). Ropemaker Place calculating an office's lifecycle footprint. *Chartered Institution of Building Services Engineer Journal*.

Berg, F., & Fuglseth, M. (2018). Life cycle assessment and historic buildings: Energy-efficiency refurbishment versus new construction in Norway. *Journal of Architectural Conservation*, 1–16. https://doi.org/10.1080/13556207.2018.1493664.

Bertram, J. (1911). *Notes on the erection of library buildings*. New York: Carnegie Corporation of New York.

Black, A. (2009). *Books, buildings and social engineering: Early public libraries in Britain from past to present*. Edited by Simon Pepper and Kaye Bagshaw. Farnham, Surrey: Ashgate.

Boughanmi, M. (2020a). 3D pdf Toxteth library. Accessed 26.6.20. https://carnegielibrariesofbritain.com/hbim-components/.

Boughanmi, M. (2020b). Life Cycle Analysis model, Toxteth library. Accessed 29.6.20. Click here to download the LCA data for Toxteth.

Brealey, R. A. (2020). *Principles of corporate finance*. Edited by Stewart C. Myers and Franklin Allen. Thirteenth edition, International student edition. New York: McGraw-Hill.

The British Newspaper Archive. Findmypast Newspaper Archive Limited and The British Library. Last modified 2017–20. Accessed 22.6.20. https://www.britishnewspaperarchive.co.uk/.

Brodie, A., Royal Institute of British Architects, British Architectural Library, Sir Banister Fletcher Library, Institute of British Architects, & Society of Architects (2001). *Directory of British architects 1834–1914*. Updated and expanded ed. London: Continuum.

The Builder (1842). Getty Research Institute. Last modified 2017–2020. Accessed 22.6.20. https://archive.org/details/getty?and%5B%5D=the+builder&sin=&sort=-addeddate.

Building News and Engineering Journal 1860–1920. Gerstein – University of Toronto. Last modified 2017–2020. Accessed 22.6.20. https://archive.org/details/buildingnewseng105londuoft.

Bull, J. W. (1993). *Life cycle costing for construction*. Blackie Academic & Professional.

Bull, J. W. (2015). *Life cycle costing: For the analysis, management and maintenance of civil engineering infrastructure*. Dunbeath, Caithness: Whittles Publishing.

Cadw. Cadw Records. n.d. Last modified 22.6.20. Accessed 22.6.20. https://cadw.gov.wales/advice-support/cof-cymru/search-cadw-records.

Caplehorn, P. (2011). *Whole life costing: A new approach*. London: Taylor & Francis.

Carnegie Corporation of New York (Author), Home Trust Company (Author), Carnegie, Andrew, 1835–1919 (Author) (1898). Carnegie Index Cards, free public libraries, outside North America. Carnegie Corporation of New York Records. Rare Book & Manuscript Library, Columbia University. Accessed 1.2.20. https://dx.doi.org/10.7916/d8-2aef-8b74.

Chartered Institute of Public Finance and Accountancy (2019). CIPFA library survey 2018/19. Accessed 5.6.20. https://www.cipfa.org/about-cipfa/press-office/latest-press-releases/decade-of-austerity-sees-30-drop-in-library-spending

Cheshire, D., & Menezes, A. (2013). *CIBSE TM54: 2013 Evaluating operational energy performance of buildings at the design stage*. The Chartered Institution of Building Services Engineers London.

CIBSE (2020). Energy Benchmarking tool dashboard. Last modified 19.6.20. https://www.cibse.org/Knowledge/Benchmarking.

Clark, J. W. (1894). *Libraries in the Medieval and Renaissance Periods. The Rede Lecture Delivered June 13, 1894*: The Project Gutenberg EBook (accessed 23.6.20).

Cole, R. J., & Kernan, P. C. (1996). Life-cycle energy use in office buildings. *Building and Environment*, 31(4), 307–317. https://doi.org/10.1016/0360-1323(96)00017-0.

Committee on Climate Change (2008). *Building a low-carbon economy: The UK's contribution to tackling climate change*. Stationery Office.

Davies, H., & Engineers Chartered Institution of Building Services (2009). *Operational ratings and display energy certificates*. London: CIBSE.

Department for Communities Buildings Database. n.d. Last modified 22.6.20. Accessed 22.6.20. https://apps.communities-ni.gov.uk/Buildings/buildMain.aspx?Accept.

Department of Finance (2020). Northern Ireland Non-Domestic Energy Performance Register. The Non-Domestic Energy Performance Certificate Register is operated by Landmark on behalf of the Department of Finance. Last modified 4.6.20. Accessed 5.6.20. https://www.epbniregisternd.com/.

Dixit, M. K., Fernández-Solís, J. L., Lavy, S, & Culp, C. H. (2010). Identification of parameters for embodied energy measurement: A literature review. *Energy and Buildings*, 42(8), 1238–1247. https://doi.org/10.1016/j.enbuild.2010.02.016. http://www.sciencedirect.com/science/article/pii/S0378778810000472.

Ed. (1902a). Library, Liverpool. *The Builder*, 12.7.1902, p. 38.

Ed. (1902b). Mr Carnegie and Liverpool free libraries. *Liverpool Echo*, Wednesday 31 December, 1902. Accessed 23.6.20. https://www.britishnewspaperarchive.co.uk/viewer/bl/0000271/19021231/048/0004.

Ed. (1902c). Mr Carnegie in Liverpool. *Liverpool Echo*, Wednesday 15 October, 1902. Accessed 23.6.20. https://www.britishnewspaperarchive.co.uk/viewer/bl/0000271/19021015/028/0005.

Ed. (1903a). Free library Kettering. *The Builder*, 25.4.1903, p. 436.

Ed. (1903b). Hammersmith public library (competition). *The Building News*, 2.10.1903. p. 445.

Ed. (1905a). Montrose public library. *The Building News*, 22.9.1905, p. 397.

Ed. (1905b). An unsolicited library. *Dundee Evening Telegraph*, Thursday 29 June, 1905. https://www.britishnewspaperarchive.co.uk/viewer/bl/0000563/19050629/140/0006.

Ed. (1917). Lincoln library. *The Builder*, 11.5.1917, p. 304.

Butcher, K. (ed.) (2012). *CIBSE Guide F (2012): Energy efficiency in buildings*. The Chartered Institution of Building Services Engineers (London).

Elefante, C. (2007). The greenest building is … one that is already built. *Forum Journal*, 21(4), 26–38.

Ellingham, I. (2006). *New generation whole-life costing: Property and construction decision-making under uncertainty*. Edited by William Fawcett. London: Taylor & Francis.

Ellingham, I. (2013). *Whole life sustainability*. Edited by William Fawcett. London: RIBA Publishing.

Energy Saving Trust. n.d. Scottish Energy Performance Certificate Register. Accessed 4.6.20. https://www.scottishepcregister.org.uk/.

Flanagan, R. (1989). *Life cycle costing: Theory and practice*. Oxford: BSP Professional.

Flanagan, R. (2004). *Whole life appraisal for construction*. Edited by Carol Jewell, Malden, MA: Blackwell Publishing.

Grimes, B. (1998). Irish Carnegie libraries: A catalogue and architectural history. http://books.google.com/books?id=Yg1UAAAAMAAJ.

Hammond, G., & Jones, C. (2011). Embodied carbon, edited by Fiona Lowrie and Peter Tse. *The inventory of carbon and energy (ICE)*.

Historic England. n.d. National Heritage List for England. Last modified 22.6.20. Accessed 22.6.20. https://historicengland.org.uk/listing/the-list/.

Historic England (2020). 2019 – Carbon in the built historic environment. Heritage Counts 2019. Accessed 22.6.20. https://historicengland.org.uk/research/heritage-counts/2019-carbon-in-built-environment/.

Historic Environment Scotland. n.d. "Listed buildings portal." Last modified 22.6.20. Accessed 22.6.20. https://www.historicenvironment.scot/advice-and-support/listing-scheduling-and-designations/listed-buildings/search-for-a-listed-building/.

Hobsbawm, E., & Ranger, T. (2012). *The invention of tradition*. New York: Cambridge University Press.

Hong, S. M. (2015). Benchmarking the energy performance of the UK non-domestic stock: a schools case study. PhD, University College London. http://discovery.ucl.ac.uk/1464471/.

Horton, R. (2020). *The COVID-19 catastrophe: What's gone wrong and how to stop it happening again*. [S.l.]: Polity Press.

Huovila, P., & United Nations Environment Programme (2007). *Buildings and climate change: Status, challenges, and opportunities*. Nairobi, Kenya: United Nations Environment Programme; Paris, France: UNEP DTIE, Sustainable Consumption and Production Branch.

Ibn-Mohammed, T., Greenough, R., Taylor, S., Ozawa-Meida, L., & Acquaye, A. (2013). Operational vs. embodied emissions in buildings—A review of current trends. *Energy and Buildings, 66*, 232–245. https://doi.org/10.1016/j.enbuild.2013.07.026. http://www.sciencedirect.com/science/article/pii/S0378778813004143.

Janda, K. B. (2011). Buildings don't use energy: People do. *Architectural Science Review, 54*(1), 15–22. https://doi.org/10.3763/asre.2009.0050.

Klinenberg, E. (2018). *Palaces for the people: How social infrastructure can help fight inequality, polarization, and the decline of civic life*. Broadway Books.

Knight, I. "iSERVcmb." Accessed 26.6.20. https://iservcmb.info/energy-conservation-opportunities/.

Langston, C. A. (2005). *Life-cost approach to building evaluation*. Oxford: Butterworth Heinemann.

Macey, F. W. (1898). *Specifications in detail*. E. & F.N. Spon: London.

Martin, R., Sunley, P., Tyler, P., & Gardiner, B. (2016). Divergent cities in post-industrial Britain. *Cambridge Journal of Regions, Economy and Society, 9*(2), 269–299. https://doi.org/10.1093/cjres/rsw005.

Menzies, G. F. (2011). *Historic Scotland technical paper 13: Embodied energy considerations for historic buildings*. Historic Scotland. https://researchportal.hw.ac.uk/en/publications/historic-scotland-technical-paper-13-embodied-energy-consideratio.

Ministry of Housing, Communities & Local Government (2019). English indices of deprivation 2019. Powered by PublishMyData ©2019 Swirrl. Data ownership and licensing as advertised on individual resources and datasets. Accessed 22.6.20. http://imd-by-postcode.opendatacommunities.org/imd/2019.

Ministry of Housing, Communities & Local Government (2020a). "Energy Performance of Buildings Data: England and Wales. All data fields other than the address and postcode data (address, address 1, address 2, address 3, postcode) available via this website are licensed under the Open Government Licence v3.0. Last modified 3.6.20. Accessed 5.6.20. https://epc.opendatacommunities.org/.

Ministry of Housing, Communities & Local Government (2020b). Energy Performance of Buildings Data: England and Wales Non-Domestic Energy Performance Register. http://www.nationalarchives.gov.uk/doc/open-government-licence/version/3/. Last modified 04/06/20. Accessed 5.6.20. https://www.ndepcregister.com/reportSearchAddressByPostcode.html https://epc.opendatacommunities.org/#register.

Ministry of Housing Communities & Local Government (2020). Energy Performance Building Certificates (EPC) in England and Wales 2008 to March 2020, Table DEC1 – at Q1 2020 (January to March), DECs in England & Wales – Number of Display Energy Certificates lodged on the Register by Energy Performance Operational Rating. Crown Copyright. Last modified 5.6.20. Accessed 5.6.20. https://www.gov.uk/government/collections/energy-performance-of-buildings-certificates.

Mohamed Abdelhalim, A., & Abouzid Sameh, F. (2011). Survey on medicinal plants and spices used in Beni-Sueif, Upper Egypt. *Journal of Ethnobiology and Ethnomedicine*, 7(1), 18. https://doi.org/10.1186/1746-4269-7-18.

Mudgal, S., Lauranson, R., Jean-Baptiste, V., Bain, J., & Broutin, N. (2009). *Study on water performance of buildings*. European Commission (DG ENV) (Paris: Publications Office of the European Union). europa.eu › environment › water › quantity › pdf ›.

News, Building (1905). *The Building News* (June 2 1905), 784.

Northern Ireland Statistics and Research Agency (2017). Northern Ireland Multiple Deprivation Measure 2017 (NIMDM2017). Accessed 10.6.20. https://deprivation.nisra.gov.uk/.

ONS (2010, revised 2013). Using Indices of Deprivation in the United Kingdom Guidance Paper, edited by the Office for National Statistics.

Paton, A., Fooks, G., Maestri, G., & Lowe, P. (2020). Submission of evidence on the disproportionate impact of COVID 19, and the UK government response, on ethnic minorities and women in the UK. Accessed 22.6.20. https://research.aston.ac.uk/en/publications/submission-of-evidence-on-the-disproportionate-impact-of-covid-19.

Prizeman, O. (2012). *Philanthropy and light: Carnegie libraries and the advent of transatlantic standards for public space*. Farnham, Surrey, England; Burlington, VT: Ashgate.

Prizeman, O. (2013). Function and decoration, tradition and invention: Carnegie libraries and their architectural messages. *Library & Information History*, 29(4), 239–257. https://doi.org/10.1179/1758348913Z.00000000046.

Prizeman, O., & Pezzica. C. 2020. GIS map Carnegie Libraries of Britain. Accessed 30.6.20. https://carnegielibrariesofbritain.com/wp-content/uploads/LCA-for-Toxteth.zip.

Prizeman, O., Pezzica, C., Taher, A., & Boughanmi, M. (2020). Networking historic environmental standards to address modern challenges for sustainable conservation in HBIM. *Applied Sciences*, 10(4), 1283. https://www.mdpi.com/2076-3417/10/4/1283.

Ramesh, T., Prakash, R. & Shukla, K. K. (2010). Life cycle energy analysis of buildings: An overview. *Energy and Buildings*, 42, 1592–1600. https://doi.org/10.1016/j.enbuild.2010.05.007.

Rye, C. (2012). The SPAB Research Report 1: U-value Report. SPAB's Building Performance Survey, supported by English Heritage. https://historicengland.org.uk/images-books/publications/spab-rr1-uvalue-report/.

Scottish Government (28 January 2020). Scottish Index of Multiple Deprivation (SIMD) 2020. Accessed 10.6.20. www.gov.scot/SIMD.

Sears, John (ed.) (1893). *The architects', surveyors' and engineers' compendium and complete catalogue*. Compendium Publishing, Co.

Smith, B. P., & Fieldsin, R. (2008). Whole-life carbon footprinting. *The Structural Engineer*, 86(6), 15–16.

Smith, Maf. (1998). *Greening the built environment*. Edited by Nicholas J. Williams and J. Whitelegg. London: Earthscan.

Taylor, C. L. (1916). Title. VI.A.3. Box 11, folder 7 Braddock, Homestead and Duquesne, 1901–1924, Spring Lake, New Jersey.

The University of Edinburgh. n.d. OS Digimap – Ordnance Survey® Collection Historic Digimap – historic maps from Landmark. Last modified 22.6.20. Accessed 1.2.20. www.digimap.edina.ac.uk.

United Nations (2015). *Transforming our world: The 2030 agenda for sustainable development*. New York.

Walker, D. M. et al. (2016). *Dictionary of Scottish architects*. St Andrews.

Xuan, W., & Hongyan, L. (2011). Energy saving and green building design of libraries: The case study of Zhengzhou library. World Library and Information Congress: 77th IFLA General Conference and Assembly, San Juan, Puerto Rico, 13–18 August 2011, IFLA. www.ifla.org › past-wlic › 196-wang-en.

Welsh Government (2020a). "Postcode lookup spreadsheet for the Welsh Index of Multiple Deprivation. Accessed 22.6.20. https://gov.wales/welsh-index-multiple-deprivation-index-guidance.

Welsh Government (2020b). Welsh Index of Multiple Deprivation (full Index update with ranks): 2019. Accessed 22.6.20. https://gov.wales/welsh-index-multiple-deprivation-full-index-update-ranks-2019.

"Welsh Newspapers." The National Library of Wales. Last modified 2017–2020. Accessed 22.6.20. https://newspapers.library.wales/.

6

CHARACTERISTICS OF GRASSLAND BURNING IN THE ASO CULTURAL LANDSCAPE, JAPAN

Yoshitaka Takahashi and Mihoko Muto

Introduction

Japan is in a heavy rainfall zone, making its climate warm and humid. Natural succession without disturbances, generally volcanic eruptions or river floods, can result in several types of vegetation reaching their climax under such conditions. For this reason, natural grasslands can hardly be seen in the Japanese archipelago, apart from limited areas of alpine or coastal grasslands where natural disturbances occur regularly. However, some traces of ancient semi-natural grasslands have been discovered in various regions throughout the country (Himiyama et al. 1995; Ogura 2012), primarily in geological regions covered with Andosol (black humic volcanic ash soil) (Shoji et al. 1999). Pedological studies on their soil profiles and genesis indicate that the grassland vegetation has survived under the monsoonal climate for a prolonged period of time (Kato 1964; Okamoto 2009). Recent studies also suggest that these grasslands are likely to be maintained by repeated fire events, since the Andosol soil contains a considerable amount of charcoal (plant carbides) particles (Yamanoi 1996; Okamoto 2009; Miyabuchi et al. 2010; Ogura 2012).

Aso, one of the most active volcanic regions on Kyushu Island, is famous for its magnificent landscape created by large-scale grasslands spread over the second largest caldera in Japan. It is one of the few remaining examples of communal grasslands, traditionally owned and managed by villages for multi-purpose use. It is also known as the second largest thatch distributor, especially for heritage conservations, in the national market today. Grasslands can be found in most ecoregions of the world; mainly located in dry or semi-arid climate zones, like savannah. In regards to the application of cultural burning, similar methods can be found in other parts of the world in terms of its purpose to increase young herbage for grazing. However, the cultural burning practiced in the Aso region (Figure 6.1) should be anticipated for its multiple aspects providing various social and environmental services other than just feeding livestock or grazing. It is rather for

DOI: 10.4324/9781003527404-8

generating a cyclical and multi-purpose use of grass resources for sustaining rural livelihoods and agricultural practices (Takahashi et al. 2017; Yokogawa & Takahashi 2017).

Figure 6.1 Cultural burning in the Aso region. Burning can eliminate shrubs but allow herbaceous plants to survive and regenerate in the next season.
Photograph by Aso Green Stock Foundation.

The traditional management system (cultural burning, mowing, and grazing), continued over thousands of years, has created a rich natural environment. It is not only remarkable to consider that these grasslands have existed for such a long time under the monsoon climate but also its ability to benefit ecosystem services; maintaining wildlife biodiversity, promoting water recharge capacity, and increasing soil carbon storage. However, annual burning is labour intensive and demands significant levels of skill. Thus, a multi-stakeholder partnership, involving collaboration among local farmers, private companies, and government agencies with the participation of volunteers, has been launched. This chapter introduces the characteristics of the grassland management system developed in the Aso region, particularly focusing on the effects and significance of cultural burning.

Distinction Between Shifting Cultivation and Cultural Burning in the Aso Region

One may consider that 'Shifting Cultivation' (also known as 'swidden farming' or 'slash and burn farming'), a widespread agricultural method practiced in many regions of the world including Japan, is a similar example to the burning practice in the Aso region. Shifting cultivation is a popular agricultural method for mountainous locations.

Forest vegetation is hewed down then set on fire to maximise the nutrients contained in burnt plants and soil for cultivating cereals, potatoes, vegetables, and other types of crops which regenerate forests in a few years' time. Among some experts and Indigenous communities, shifting cultivation is recognised as a sustainable agricultural method for rotational agroforestry since it makes the soil rich enough to eliminate artificial fertilisers, herbicides, and pesticides if it is for short-term cultivation. However, there is a predominant view on the shifting cultivation as it can often degrade soil fertility and wildlife habitat; particularly when the fallow period is too short (Cherrier et al. 2018). In many countries, national governments have discouraged shifting cultivations in upland areas due to their predominantly negative perceptions of ecological impacts, and more generally, setting fire itself is regulated as an environmentally destructive activity which sometimes can cause soil erosion and uncontrolled forest fires (Cherrier et al. 2018; Forsyth & Walker 2008; Friederichsen and Neef 2010). However, cultural burning in the Aso region is well prescribed and quite different from other similar methods as it maximises sustainable use of grass resources.

In the case of slash-and-burn farming, for example, a large volume of wood fuel and longer burning time can result in the ground surface temperature reaching several hundred degrees, and the temperatures at 2–5 cm below the surface can range between several ten to hundred degrees. This highly intensive flame can result in killing many plant seeds, rhizomes, soil-dwelling insects, as well as weeds and harmful insects (Shindo et al. 1988; Su et al. 1995; Suttie et al. 2005). In contrast to slash-and-burn farming, the flame on grasslands burns at low intensity and shifts quickly; therefore, the ground does not reach relatively high temperature during the burning process (Iwanami 1969; Tsuda 2009), which allows many herbaceous plants, insects, and small animals to survive underground thus maintaining a high level of biodiversity.

Additionally, the focus of cultural burning in the case of Aso grasslands is brush control (eliminating shrubs and trees), which enables grass to regrow and maintain grass resources. On this point, the burning practice in the Aso region is quite different from shifting cultivations, it is similar to the burning practice in dry or semi-arid climate zones such as savannah, Central Asia, South Africa, and Australia, etc. (Suttie et al. 2005; Murphy & Bowman 2007). In those latter regions, cultural burning is practiced to increase herbaceous plants which are crucial when grazing livestock,[1] however as mentioned earlier, the burning practice in the Aso grasslands is not only for livestock husbandry but to maximise grass resources for many different purposes. Thus, there is a distinct difference between the cultural burning in the Aso region and other similar methods practiced in other parts of the world when considering the purpose of maintaining grasslands.

Cultural burning is a powerful tool in grassland management that can remove unpalatable scrubs and woods by promoting regrowth of

[1] The Aboriginal Australians, who are Indigenous hunters, also practice cultural burning in grasslands to increase the number of herbivores, such as kangaroos for them to hunt (Murphy & Bowman 2007). The Native Americans also burn forests and grasslands, which is believed to encourage plant growth and increase the number of deer and bison (Hobbs & Spowart 1984).

young grasses which is suitable for grazing (Suttie et al. 2005). On the other hand, the enforcement of burning practice must carefully be controlled and timed. Prescribing and controlling fire is a difficult and labour-intensive task, however only substantial preparation can avoid tremendous damage to wildlife. For this reason, many developed countries have regulations for application of prescribed fire to natural vegetation. For example, *Calluna vulgaris*-dominated upland heathland in the United Kingdom has adopted annual rotations in burning, which creates patches on the heathland. The burning season is strictly limited by law (The Muirburn Code) to minimise the impact on wildlife (Scottish Executive 2011; Suttie et al. 2005).

In the case of the Aso region, regular cultural burning has been prescribed under the democratic governance of local communities. It is a 'wise choice' to enhance biodiversity and allow for sustainable use of grass resources at the same time. Furthermore, as mentioned above, wildlife conservation is another distinctive feature of grassland restoration in the case of the Aso region. Public support, triggered by re-evaluating traditional management systems, have resulted in initiating partnerships involving various stakeholders. Under these circumstances, regular and controlled burning practices can give a potential to positive effects on the protection of grassland, as well as the preservation of endangered species (Koyama et al. 2017; Takahashi et al. 2017).

Distinctive Features of Grassland Burning in the Aso Region

In comparison to similar burning practices found in other countries, the following features are distinguished in the case of Aso grasslands.

Meteorological Conditions

In many countries located in arid or semi-arid climate zones, such as savannah, Central Asia, and North or South America, it is relatively easy to maintain grassland vegetation. For example, cultural burning is required only once in several years or decades in the prairies of North America, depending on grass fuel load and rainfall, to keep the grassland as an open field (Robinett 1994; Sargent & Carter 1999; Fuhlendorf et al. 2009). It is almost the same cycle for the savanna in Africa (Siegfried 1981; Oluwole et al. 2008; Trollope & Trollope 2004; Hobbs & Spowart 1984).

On the contrary, the climax vegetation of natural succession in Asian monsoon countries, including Japan, is 'forest' which means that the grasslands are eventually replaced with broad-leaved forests if there are no disturbances. For this reason, human interventions such as regular burning practices are essential to keep the grassland as an open field. To achieve this, solid resource governance of local communities is a critical part in the grassland management system. Therefore, the magnificent landscape with large-scale grasslands can be appreciated as an accumulation of human efforts. It has been achieved by burning, mowing, and grazing continued for over 10,000 years, in spite of its nature that grasslands can easily be predominated by succession (Miyabuchi & Sugiyama 2006; Miyabuchi et al. 2010).

Purpose of Using Grasses

Second, there is a great difference in the purpose of burning practices. In many other countries or regions, it is applied to stimulate the growth of grass for grazing. However, in the case of the Aso region, grasslands maintained by cultural burning can offer multiple benefits to all sorts of rural livelihoods and agricultural practices, hence it plays an essential role in human settlement. Although the volcanic ash soil (Andosol), which covers wide areas of the Aso region, is not originally suitable for cultivation, its productivity can be improved by mixing organic matters such as green manure, compost, or barnyard manure. For this reason, farmers living in the Aso region use grass as a fertiliser for cultivating paddy field (or dry field), in which case, feeding livestock is only one part of their cyclical agricultural practices (Takahashi et al. 2017; Yokogawa & Takahashi 2017). Cows and horses are used as a work force to support farming, transportation, as well as for provisioning of manure; they have not been raised for meat consumptions until recent years (Figure 6.2). Grass can offer a variety of products for human needs; green manure, compost, and barnyard manure, as well as roofing thatch and fuels. The grass gathered in native grasslands is a vital source for regional and cyclical agricultural practices, thus cultural burning has become a core activity to generate the entire agricultural system developed in the Aso region (Figure 6.3).

Figure 6.2 Cows and horses used to be important work force for agricultural practices until the 1960s.
Photograph by Norio Otaki.

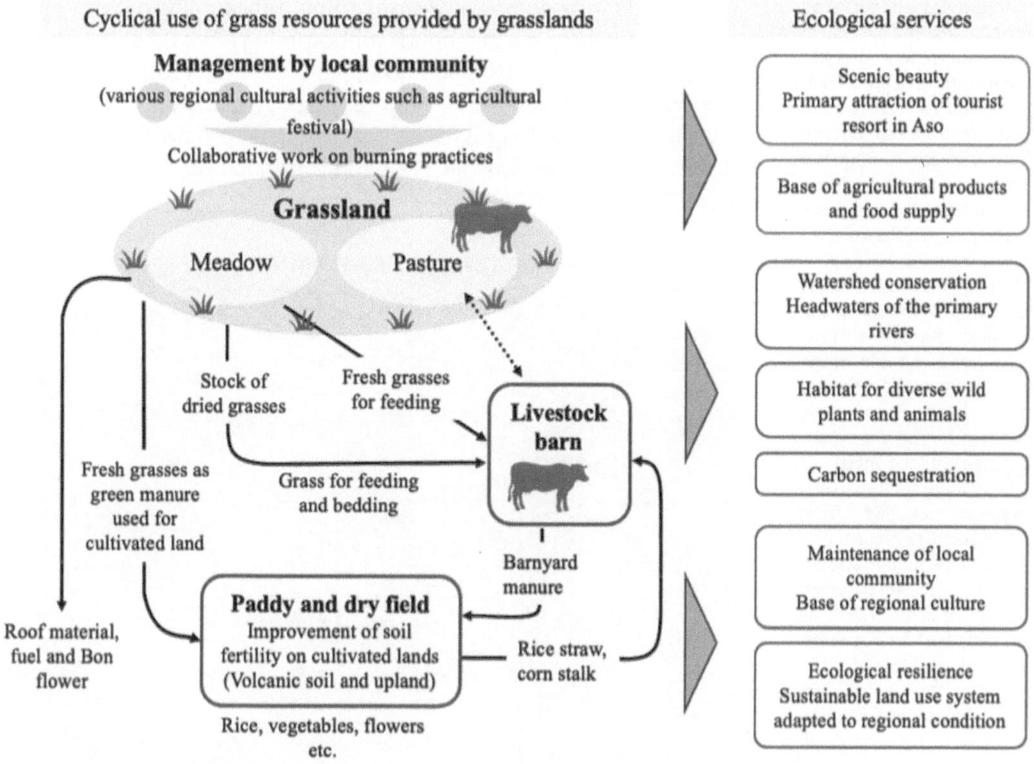

Cyclical use of grass resources provided by grasslands   Ecological services

Management by local community
(various regional cultural activities such as agricultural festival)
Collaborative work on burning practices

Grassland

Meadow   Pasture

Stock of dried grasses   Fresh grasses for feeding

Fresh grasses as green manure used for cultivated land

Grass for feeding and bedding

Livestock barn

Barnyard manure

Roof material, fuel and Bon flower

Paddy and dry field
Improvement of soil fertility on cultivated lands (Volcanic soil and upland)

Rice straw, corn stalk

Rice, vegetables, flowers etc.

Scenic beauty
Primary attraction of tourist resort in Aso

Base of agricultural products and food supply

Watershed conservation
Headwaters of the primary rivers

Habitat for diverse wild plants and animals

Carbon sequestration

Maintenance of local community
Base of regional culture

Ecological resilience
Sustainable land use system adapted to regional condition

Figure 6.3 Cyclical use of grass resources and other services provided by grasslands.
Illustrated by the author.

The diversity in grassland utilisation has far more than a single value which provides a wide range of benefits to eco services. It also has created unique traditions, culture, and seasonal landscapes such as a custom of arranging wildflowers for Bon Festival (*Bon flower*),[2] dried

[2] Bon Festival is a local custom that takes place every August, the graves of ancestors are decorated with wild flowers 'Bon Bana' (Bon flowers) for the festival.

Figure 6.4 Cows carrying down grass from hay meadows along *Kusa-no-michi* in the 1960s.
Photograph by Norio Otaki.

[3] Cut grass dried, gathered into bundles, and piled up in small heaps. It is a traditional method of preserving grass, and is used to feed cattle and spread on the floors of barns during the winter, which are finally applied to cultivation fields as compost. This is a dialect mainly from the Aso region.

[4] Huts made of silver grass for encampment near the hay fields during the mowing period in autumn. In the Aso region, it was practiced in the Hatabe Plain on the northern rim of the mountain area until the 1960s.

[5] A boundary created to distinguish meadows from grazing land, to prevent cattle from escaping, and to increase efficiency when mowing. It is often used as the boundary for the rightful use of grazing land or the boundary between villages. The total length is estimated as long as 500 km in the Aso region.

[6] Cattle tracks marked along a contour line by hoof pressure as grazing cattle horizontally on the caldera wall over a long period of time.

[7] Cattle tracks created on the caldera wall connecting the back of each village and the grassland (meadow) located on the outer rim about 500 m above the caldera floor. In the 1950s, the footpaths were used as many as 200 times per household, and were once a lifeline for farmers (Yokogawa & Takahashi 2017).

grass stacking after mowing (*Kusakozumi*),[3] encamping huts for mowing (*Kusadomari*),[4] earth mounds for bordering (*Tomo*),[5] paths for cattle to graze grass (*Ushi-no-michi*),[6] and vertical footpaths on the caldera wall (*Kusa-no-michi*)[7] (Figures 6.4, 6.5, and 6.6). This rural lifestyle associated with grasslands has developed a basis of agricultural rituals and festivals which have been continued for several hundred years (Kajihara 2016).

Figure 6.5 Camping at *Kusadomari* in the 1960s.
Photograph by Norio Otaki.

Soil Carbon Sequestration

Cultural burning has another significance in creating a soil carbon (C) storage (or sequestration) in the Aso grasslands. Andosol (black volcanic ash) is a common profile for volcanic regions in the Japanese

Figure 6.6 *Kusakozumi* **(dried grass stack) after mowing.** Photograph by Norio Otaki.

archipelago; the Aso region is no exception. In many cases, this type of soil is accumulated within the volcanic tephra underneath the grassland vegetation, which has stayed in place for a prolonged period of time. It contains a considerable amount of carbon, resulting in serving as a carbon sink. The most part of this organic carbon is supplied by the roots and rhizomes of dominant grass species, *Miscanthus sinensis* and *Pleioblastus chino* var. *viridis*, as well as the charcoals remaining after the burning process (Toma et al. 2010, 2013) (Figure 6.7). Recent studies on phytolith (or plant opal) and charcoal fragments contained in the soil verify that the grasslands have been maintained and extended across the region in response to frequent fire events for more than 10,000 years (Miyabuchi & Sugiyama 2006; Miyabuchi et al. 2010). It also suggests that the carbon has constantly been accumulating over a prolonged period.

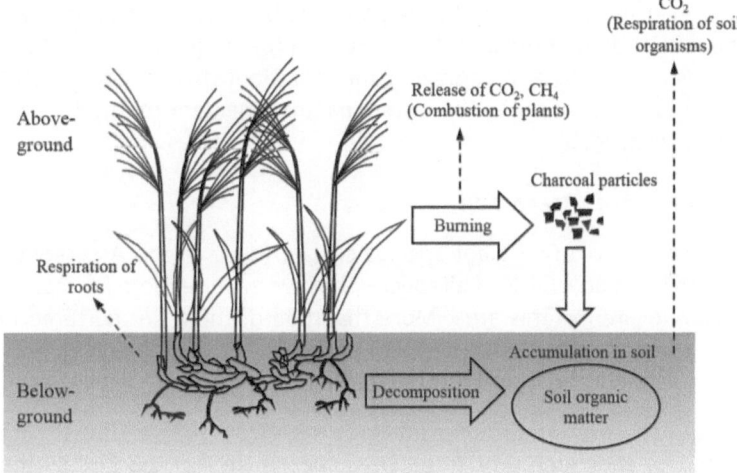

Figure 6.7 Schematic diagram of soil carbon sequestration. Illustrated by the author.

[8] The number of house-
holds in the seven munic-
ipalities of Aso region is
24,542 (according to the
2015 National Tax Sur-
vey); carbon emissions
per household in 2015
(2005) were 4.92 t (http://
www-gio.nies.go.jp/
aboutghg/nir/nir-j.html).

The total carbon stock of the grassland is 232 t C/ha (Toma et al. 2013), and the annual sequestration rate of soil C is equivalent to 6.9 t CO_2 /ha (Toma et al. 2013), which makes the Aso region as one of the largest soil carbon stocks in the world. The total area of burnt grass-land (15,887 ha, Kumamoto Prefecture 2022) can absorb a carbon amount equivalent to CO_2 emissions from 28,800 households, or 121% of the emissions of total households inhabited in the Aso region[8] if the latter value is applied. Because the flame shifts swiftly and burns out quickly, it does not significantly reduce the soil C nor increase CO_2 and methane (CH_4) emissions from the ground (Toma et al. 2010).

Thus, the annual burning practice in the grass-dominated vegetation community plays a significant role in enhancing C storage of the ground, stability of regrowth, and productivity of grass plants. More-over, despite the continual use of aboveground biomass without using any soil tillage and fertiliser, the grasslands can continue to increase soil C input (Howlett et al. 2013; Toma et al. 2013). It is an outstanding exam-ple of a combination of natural and human works which has been ongo-ing for more than 10,000 years in the Aso grasslands (Miyabuchi et al. 2010; Ogura 2012).

Several studies have also recorded that some burning practices would decrease soil organic carbons (e.g. Fernández et al. 1997; Novara et al. 2011; Cheng et al. 2013; Abdalla et al. 2016), whereas the grasslands in the Aso region have succeeded in accumulating below-ground C despite the successive burning practices (Toma et al. 2010). It means that the long tradition of cultural burning and proactive utilisation of grass resources have not affected land degradation, such as severe soil erosion or organic carbon loss. In other words, a highly sustainable agricultural practice has been achieved through regular maintenance of grasslands. Such long-term and continual applications of cultural burning can barely be seen in any other part of the world except the Aso region.

According to a recent study carried out on C sequestration of soil (Toma et al. 2012), *Miscanthus sinensis* grasslands, such as those widely distributed in the Aso region, may be considered to be more beneficial than afforesting them. Even in the future, annual cultural burning, mowing, and/or grazing, may contribute to mitigating global warming by continuing the traditional management method (Toma et al. 2012, 2016).

Biodiversity Conservation

An estimated 1,600 plant species can be found in the Aso region, which sums up to 70% of all species that live in Kumamoto Prefecture being located in this area. More than 600 of them are nurtured in grasslands, and many species of birds and butterflies share the wood-ed-grassland environment (Kumamoto Pref. 1998).

Of the approximately 600 plants grown in grasslands, 72 of them are listed as endangered or threatened species on the Red List by the

Ministry of Environment (Takahashi et al. 2017). In total, about 10% of grassland plant species living in the Aso region are facing the risk of extinction, which makes the grasslands one of the highly concentrated refuges for Japanese endangered species. These grassland plants and animals have been nurtured in grassland vegetation communities, as a consequence of human activities such as cultural burning.

In terms of biodiversity, many tertiary relict species, *Viola orientalis*, *Echinops setifer*, *Silene sieboldii*, and *Campanula glomerata* var. *dahurica* (which had migrated southward during the glacial age when Kyushu Island was attached to the Asian continent), are concentrated in the Aso grasslands (Okubo 2002; Takahashi et al. 2017). Such relict species illuminate the characteristics of flora and fauna of the region (Uchino 2016). As a result of global warming during the post-glacial period, some of these species had died out, while other species were forced to reduce their habitat. However, many of these species had taken advantages of grassland vegetation, thus expanding their habitual territory. These man-made grasslands have consequently resulted in creating an ideal refuge for wildlife which promoted the preservation of flora and fauna during the ice ages, as well as the post-glacial re-colonisation of these species (Uchino 2016; Takahashi et al. 2017). In other words, the long tradition of human interventions to the grasslands has resulted in protecting biodiversity from global warming in the case of the Aso region.

Water Recharge Capacity

The widely covered areas of grassland play a vital role in creating natural groundwater storage in the Aso region. The leaf-stem blades and roots of grass can slow the rainwater flow and allow water to infiltrate underground. The stored underground water eventually returns to the surface as a water source for living creatures aboveground.

In terms of water budget of vegetation, forests have more 'interception losses' through the tree canopy and 'transpiration' from leaf stomata than any other types of vegetation (Baumgartner 1967; Fahey & Jackson 1997; Kubota 2004). On the other hand, grasslands have a lower ratio of evapotranspiration than the ones of forests or cultivated land in comparison to annual precipitation, therefore a large amount of water can reach the ground surface (Fahey & Jackson 1997; Kubota 2004; Shimatani 2022). In terms of the infiltration capacity from the ground surface to underground, broad-leaved forests have the highest values; however, grasslands also show considerably high values according to research (Murai 1993).

These studies indicate the high capacity of grasslands in groundwater recharge. In addition to this, a study carried out at the western foot of Mt. Aso shows an interesting result in which the groundwater recharge rate of the grassland catchment is about 30% higher than the forest catchment according to two different hydrological methods; the water budget method and the Displacement Flow Model (Figure 6.8, Kudo et al. 2012; Kudo et al. 2016). If the grasslands are maintained properly the Aso region, which originally has a large underground water storage due to highly permeable soil (Andosol) and strata (pyroclastic flow deposit) (Mushiake et al. 1981; Kudo et al. 2016), can offer a large potential for water recharge.

Figure 6.8 Comparison of water recharge capacities between forests and grassland. Modified from Kudo et al. (2012).

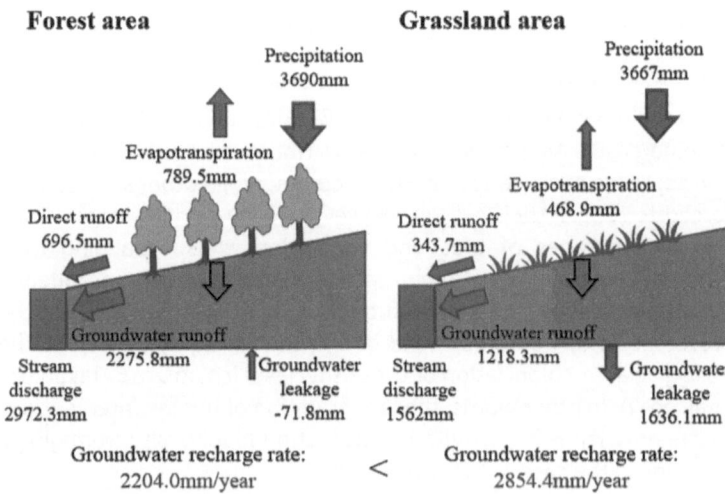

With an annual rainfall of over 3,000 mm, the Aso region can charge a large amount of rainwater, which makes the region the source of approximately 1,500 natural springs and six national first-class rivers. The Aso region is the headwater of six major rivers, including the largest river on Kyushu Island, Chikugo River, enriching the lives of 5 million residents living in the watersheds. For this reason, the Aso region is called as 'the water reservoir of Kyushu Island' (Yokokawa & Takahashi 2017).

Disaster Prevention and Mitigation

The Aso region is located in the second largest caldera in Japan, and centred on active volcanos. Its geographical and geological conditions make the whole region extremely undesirable for living. This is mainly due to 1) unstable ground conditions created by relatively recent volcanic activities; 2) surrounded by steep slopes of the outer rim which frequently cause landslides after rainfall, strong winds, and earthquakes; 3) volcanic ash soil (Andosol) with low bulk density and soft mass deposited on impervious bedrock; 4) the loosely bounded soil particles; and 5) high annual precipitation due to its monsoon climate.

Therefore, if the grasslands were abandoned or afforested, a large volume of aboveground biomass (trees) containing considerable amounts of moisture can increase the risk of hazardous events like landslides (Kajihara 2016). By contrast, if grasslands are maintained properly, they can contribute to mitigating disasters by their small amount of aboveground biomass retention, which allows rainfall to run off the ground surface and/or smoothly infiltrate underground. In addition to this, when a landslide occurs, it has been observed that destruction usually stops at the topmost scars in past disaster sites in the Aso grasslands (Matsushi et al. 2013). It is commonly recognised that burnt grasslands can contribute to mitigating the scale of landslides, and consequently contribute to preserving its landscape with ecological stability in the case of the Aso region (Kajihara 2016; Yokogawa & Takahashi 2017).

Hence, long-term preservation of the landscape on such an extremely unstable site are pretty much dependent on human interventions, such as cultural burning for instance. Cessation of annual burning can also lead to a build-up of undesirable combustible materials (Figures 6.9 and 6.10), which can cause highly-intensive and destructive fire events if ignited (Yokogawa & Takahashi 2017). Disaster prevention and mitigation are other unique functions of Aso grasslands which are barely seen in other parts of the world.

Contemporary Values of the Grassland Management System

The traditional grassland management system based on *Iriai* (commonage or common land) has worked effectively for inhibiting post-war overdevelopment. The public nature and autonomy of this custom should be re-evaluated today, owing to a rising public consciousness in regards to current global issues such as biodiversity conservation and environmental protection (Yokogawa & Takahashi 2017). The well-maintained grassland is also a core element in the scenic beauty of Aso-Kuju National Park. It is a remarkable example which demonstrates a successful integration between overdevelopment control against negative impacts and the conservation of cultural landscapes created by agricultural practices, if administrative bodies are highly motivated and cooperative with environmental conservation.

Moreover, re-structuring the grassland management system is also a crucial issue. Socio-economic changes, especially labour shortages and population ageing of local agricultural communities, will be a key

**Figure 6.9
Transformation of
the landscape in
abandoned grassland.
Burnt grassland in May
around 1980.**
Photograph by
Norio Otaki.

**Figure 6.10 Same place
in May 2013, where
burning has been
discontinued for more
than 30 years.**
Photograph by Yoshitaka
Takahashi.

element when justifying future operations. In correspondence to these issues, which the subject can be shared with other developed countries, a cooperative framework has been organised to continue the current practice. The initiative is an outcome of collaborations between local farmers and volunteers who are largely organised and support *noyaki* (cultural burning) and *wachikiri* (preparation of firebreaks) (Figure 6.11). A variety of funds, human resources, and programmes including CSR (corporate social responsibility) activities by local companies, the Aso Grassland Restoration Fundraising Campaign, government schemes and subsidies, are also available to reinforce their activities (Figure 6.12, Takahashi et al. 2017).

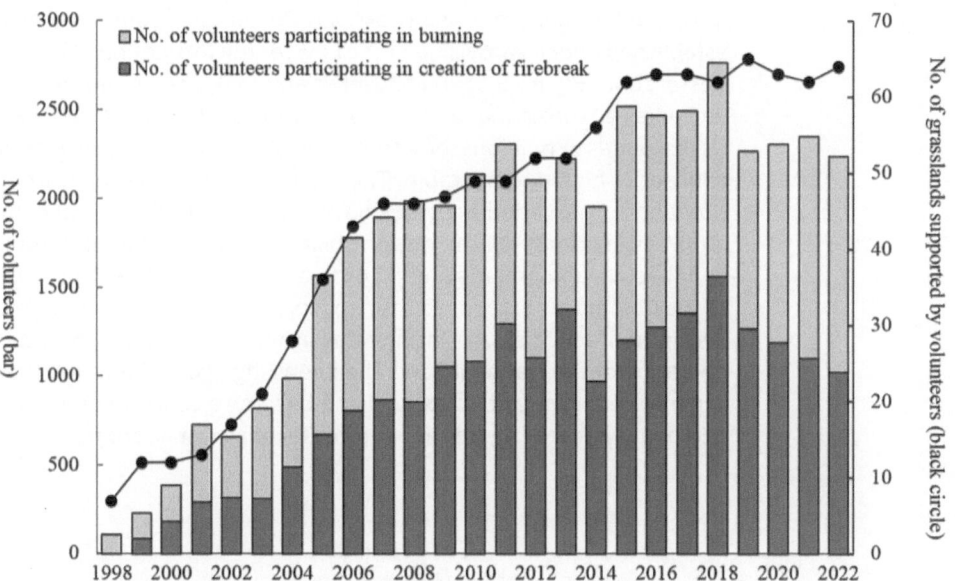

Figure 6.11 Summary of the numbers of volunteers participating in *Noyaki* (cultural burning) and *Wachikiri* (creation of mown firebreaks).
Illustrated by the author, based on annual figures released by Aso Greenstock Foundation, Japan 2023.

Utilization of Aso Grassland Restoration Fund
Total donation amount: 149.8 million yen (from 2011 to 2019)

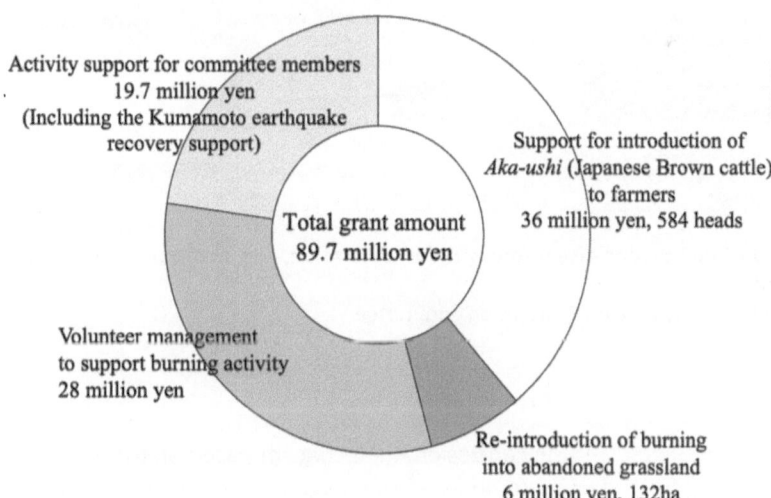

Figure 6.12 Breakdown of the total expenditure of Aso Grassland Restoration Fund. Total amount of donation: 149.8 million yen (from 2011 to 2019).
Illustrated by the author, based on figures released by Aso Grassland Restoration Committee, Japan 2020.

In December 2005, the Aso Grassland Restoration Committee was established under a provision of the Law for the Promotion of Nature Restoration (Figure 6.13, Takahashi et al. 2017). The Committee is a partnership consisting of various members, who have different backgrounds or opinions; however, their ability to develop an overall concept is highly remarkable. The leaders responsible for restoration and conservation activities have firm roots in their local communities, by clarifying the responsibility of each member according to collective agreements and co-ordinate their activities. The scope of the Committee extends to the Aso-Kuju National Park, where their restoration activities are inseparable from the administration and operation of the National Park. The knowledge gained through their experience is expected to expand and become a concrete park management system in the coming future (Takahashi et al. 2017).

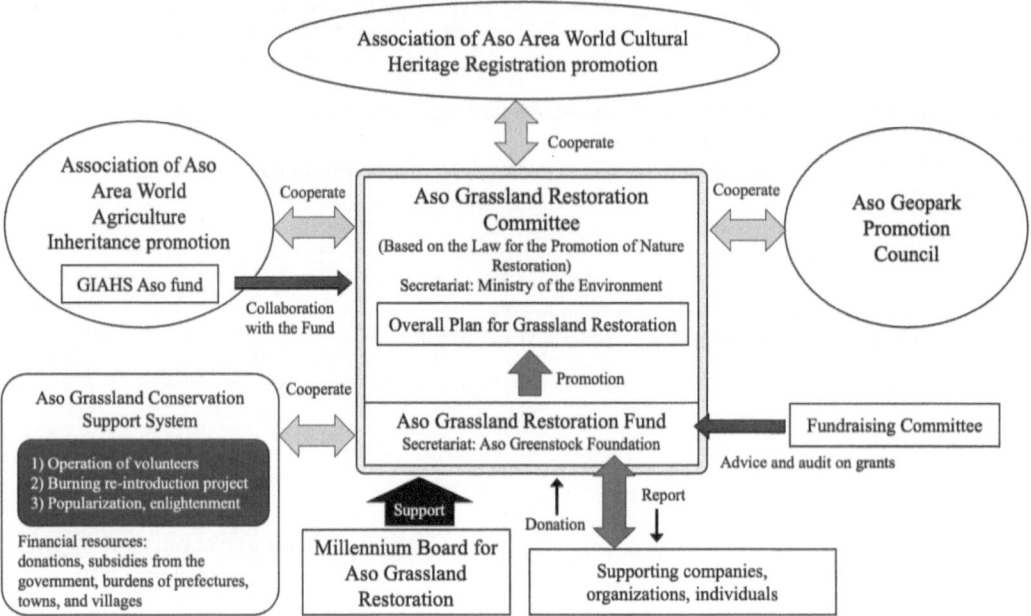

Figure 6.13 The collaborative framework for restoration, conservation, and promotion of the grasslands.
Illustrated by Aso Grassland Restoration Committee.

Conclusion

In conclusion, their concept based on the idea for the grasslands as a common-pool resource, which is firmly embedded into social structure by involving public-private partnerships, can certainly attract various stakeholders beyond the national boundary (Takahashi et al. 2017). The cultural landscape and its management developed in the Aso region should raise a wider public consciousness to the traditional form of agricultural practice, which symbolises a contemporary local-common

management system for multi-faceted aspects of social and environmental services, especially in developed countries where socio-economic changes have affected global issues.

Reference List

Abdalla, K., Chivenge, P., Everson, C., Mathieu, O., Thevenot, M., & Chaplot,
V. (2016). Long-term annual burning of grassland increases CO_2 emissions
from soils. *Geoderma, 282,* 80–86.

Baumgartner, A. (1967). *Energetic base for differential vaporization from forest
and agricultural lands,* Proceedings of the International Symposium on
Forest Hydrology, Pergamon Press, 381–389.

Cheng, C. H., Chen, Y. S., Huang, Y. H., Chiou, C. R., Lin, C. C., & Menyailo,
O. V. (2013). Effects of repeated fires on ecosystem C and N stocks along a
fire induced forest/grassland gradient. *Journal of Geophysical Research:
Biogeosciences, 118*(1), 215–225.

Cherrier, J., Maharjan, S. K., & Maharjan, K. L. (2018). Shifting cultivation:
Misconception of the Asian governments. *Journal of International Development and Cooperation, 24*(2), 71–82.

Fahey, B., & Jackson, R. (1997). Hydrological impacts of converting native forests and grasslands to pine plantations, South Island, New Zealand. *Agricultural and Forest Meteorology, 84,* 69–82.

Fernández, I., Cabaneiro, A., & Carballas, T. (1997). Organic matter changes
immediately after a wildfire in an Atlantic forest soil and comparison with
laboratory soil heating. *Soil Biology and Biochemistry, 29*(1), 1–11.

Forsyth, T., & Walker, A. (2008). *Forest guardians, forest destroyers: The politics
of environmental knowledge in Northern Thailand, Chiang Mai.* University of
Washington Press, 1–304.

Friederichsen, R., & Neef, A. (2010). Variations of late socialist development:
Integration and marginalization in the northern uplands of Vietnam and
Laos. *European Journal of Development Research, 22*(4), 564–581.

Fuhlendorf, S. D., Engle, D. M., Kerby, J., & Hamilton, R. (2009). Pyric her-
bivory: Rewilding landscapes through the recoupling of fire and grazing.
Conservation Biology, 23, 588–598. Available at: https://doi.org/10.1111/j.
1523-1739.2008.01139.x

Himiyama, Y., Arai, T., Ota, I., Kubo, S., Tamura, T., Nogami, M., Murayama, Y.,
& Yorifuji, T. (1995). *Atlas: Environmental change in modern Japan.* Tokyo:
Asakura Publishing, 1–187. (In Japanese with English tables.)

Hobbs, N. T., & Spowart, R. A. (1984). Effects of prescribed fire on nutrition of
mountain sheep and mule deer during winter and spring. *Journal of Wildlife
Management, 48*(2), 551–560.

Howlett, D. S., Toma, Y., Wang, H., Sugiyama, S., Yamada, T., Nishiwaki, A.,
Fernandez, F., & Stewart, J. R. (2013). Soil carbon source and accumulation
over 12,000 years in a semi-natural *Miscanthus sinensis* grassland in southern Japan. *Catena, 104,* 127–135. Available at: https://doi.org/10.1016/j.
catena.2012.11.002

Iwanami, Y. (1969). Temperatures during Miscanthus type grassland fire and
their effects on the regeneration of *Miscanthus sinensis. The Reports of the
Institute for Agricultural Research Tohoku University, 20,* 47–88.

Kajihara, H. (2016). Life and ethnic. *Survey report of Aso cultural landscape – vol.
detailed investigations –* (Council for Aso Area World Culture Heritage ed.),
Aso: Council for Aso Area World Culture Heritage, 302–307. (In Japanese.)

Kato, Y. (1964). Some problems on the genesis of 'Kuroboku' soils. *The
Quaternary Research, 3*(4), 212–222. (In Japanese with English summary.)

Koyama, A., Koyanagi, T. F., Akasaka, M., Takada, M., & Okabe, K. (2017). Combined burning and mowing for restoration of abandoned semi-natural grassland, *Applied Vegetation Science*, *20*(1), 40–49.

Kubota, J. (2004). Forest and water – reality and myth. *Kagaku*, *74*, 311–316. (In Japanese.)

Kudo, K., Shimada, J., & Tanaka, N. (2012). *The estimation of groundwater recharge rate for different land use – observation study at paired forest and grassland catchments. Proceedings of 39th IAH Congress*, *419*, 1–10.

Kudo, K., Shimada, J., Maruyama, A., & Tanaka, N. (2016). The quantitative evaluation of groundwater recharge rate using Displacement Flow Model with stable isotope ratio in the soil water of difference vegetation. *Journal of Groundwater Hydrology*, *58*(1), 31–45. (In Japanese with English summary.)

Kumamoto Prefecture (1998). *Wild species of fauna and flora in Kumamoto Prefecture that are important for protection: Red Data Book Kumamoto*. Kumamoto: Environment Conservation Section of Kumamoto Prefectural Government, 1–381. (In Japanese.)

Kumamoto Prefecture (2022). *Report of Aso Grassland Maintenance and Restoration Basic Survey*. Available at: https://www.pref.kumamoto.jp/uploaded/life/181104_443400_misc.pdf [Accessed 2 May 2024].

Matsushi, Y., Saito, H., Fukuoka, H., & Furuya, G. (2013). Landslides of tephra deposits on hillslope of the Aso caldera wall and volcanic central cones by the North-Kyushu heavy rainfall at July 2012. *Annuals of Disaster Prevention Research Institute*, *56B*, 237–241. (In Japanese with English summary.)

Miyabuchi, Y., & Sugiyama, S. (2006). A 30,000-year phytolith record of a tephra sequence, east of Aso caldera, southwestern Japan. *The Quaternary Research*, *45*(1),15–28. (In Japanese with English summary.)

Miyabuchi, Y., Sugiyama, S., & Sasaki, N. (2010). Phytolith and macroscopic charcoal analyses of the Senchomuta drilling core in Asodani Valley, northern part of Aso caldera, Japan. *Journal of Geography*, *119*(1), 17–32. (In Japanese with English summary.)

Murai, H. (1993). Comparison on hydrological characteristics with broad leaved forest, needle leaved forest and grassy land. *Water Science*, *37*, 1–40. (In Japanese.)

Murphy, B. P., & Bowman, M. J. S. (2007). The interdependence of fire, grass, kangaroos and Australian Aborigines: A case study from central Arnhem land, northern Australia. *Journal of Biogeography*, *34*(2), 237–250. Available at: https://doi.org/10.1111/j.1365-2699.2006.01591.x

Mushiake, K., Takahashi, Y., & Ando, Y. (1981). Effects of basin geology on river-flow regime in mountainous area of Japan. *Proceedings of the Japan Society of Civil Engineers*, *309*, 51–62. (In Japanese.)

Novara A, Gristina, L., Bodì, B. M., & Cerdà, A. (2011). The impact of fire on redistribution of soil organic matter on a Mediterranean hillslope under maquia vegetation type. *Land Degradation & Development*, *22*(6), 530–536.

Ogura, J. (2012). *History of radically changed vegetation of Japan*, Tokyo: Kokin-Shoin, 1–343. (In Japanese.)

Okamoto, T. (2009). Charcoal in forest soil as an indicator of past fires. *Shinrin Kagaku*, *55*, 18–23. (In Japanese.)

Okubo, K. (2002). The present state in the study of biological diversity on semi-natural grassland in Japan. *Japanese Journal of Grassland Science*, *48*, 268–276. (In Japanese.)

Oluwole, F. A., Sambo, J. M., & Sikhalazo, D. (2008). Long-term effects of different burning frequencies on the dry savannah grassland in South Africa. *African Journal of Agricultural Research*, *3*, 147–153.

Robinett, D. (1994). Fire effects on southeastern Arizona plains grassland. *Rangelands*, 16(4), 143–148.

Sargent, M. S., & Carter, K. S. (ed.) (1999). *Managing Michigan wildlife: A landowners guide*. East Lansing: Michigan United Conservation Clubs, 1–297.

Scottish Executive (2011). *The Muirburn Code*. Available at: https://www.nature.scot/professional-advice/land-and-sea-management/managing-land/upland-and-moorland/muirburn-code [Accessed 2 May 2024].

Shimatani, K. (2022). The value of Miscanthus-type grassland as a water resource. *National Parks*, 803, 6–8. (In Japanese.)

Shindo, T., Sugawara, S., & Ueki, K. (1988). Studies in cropping system of slash-and-burn method of agriculture and its origin. XX. Effects of burning on weed restrain. *Japanese Journal of Farm Work Research*, 23, 111–116. (In Japanese.)

Shoji, A., Suyama, T., & Sasaki, H. (1999). *Distribution and site condition of semi-natural grassland in Japan*, Proceedings from the *6th International Rangeland Congress*, 312–313.

Siegfried, W. R. (1981). The incidence of veld-fire in the Etosha National Park, 1970–1979. *Madoqua*, 12, 225–230.

Su, J., Katagiri, S., Kaneko, N., & Nagayama, Y. (1995). Changes of soil temperature and ashing extent of organic matter during burning of forest floor. *Transactions of Kansai Branch of the Japanese Forestry Society*, 4, 53–54. (In Japanese with English summary.)

Suttie, J. M., Reynolds, S. G., & Batello, C. (ed.) (2005). *Grassland of the world* (*Plant Production and Protection Series* No. 34). Rome: Food and Agriculture Organization of the United Nations (FAO), 1–514.

Takahashi, Y., Neef, A., & Yokogawa, H. (2017). Conservation and restoration of traditional grasslands in the Mount Aso Region, Kyushu, Japan: The role of collaborative management and public policy support. Shifting Cultivation Policies – Balancing Environmental and Social Sustainability – (Cairns, M. F. ed.). *CAB International*, 174–192.

Toma, Y., Fernandez, F. G., Nishiwaki, A., Yamada, T., Bollero, G., & Stewart, J. R. (2010). Aboveground plant biomass, carbon, and nitrogen dynamics before and after burning in a seminatural grassland of *Miscanthus sinensis* in Kumamoto, Japan. *Global Change Biology Bioenergy*, 2(2), 52–62. Available at: https://doi.org/10.1111/j.1757-1707.2010.01039.x

Toma, Y., Armstrong, K., Stewart, J. R., Yamada, T., Nishiwaki, A., & Fernández. F. G. (2012). Carbon sequestration in soil in a semi-natural *Miscanthus sinensis* grassland and *Cryptomeria japonica* forest plantation in Aso, Kumamoto, Japan. *Global Change Biology Bioenergy*, 4(5), 566–575. Available at: https://doi.org/10.1111/j.1757-1707.2012.01160.x

Toma, Y., Clifton-Brown, J., Sugiyama, S., Nakaboh, M., Hatano, R., Fernández, F. G., Stewart, J. R., Nishiwaki, A., & Yamada, T. (2013). Soil carbon stocks and carbon sequestration rates in seminatural grassland in Aso region, Kumamoto, Southern Japan. *Global Change Biology*, 19(6), 1676–1687. Available at: https://doi.org/10.1111/gcb.12189

Toma, Y., Yamada, T., Fernández, F. G., Nishiwaki, A., Hatano, R., & Stewart, J. R. (2016). Evaluation of greenhouse gas emissions in a *Miscanthus sinensis* Andersson-dominated semi-natural grassland in Kumamoto, Japan. *Soil Science and Plant Nutrition*, 62(1), 80–89. Available at: https://doi.org/10.1080/00380768.2015.1117944

Trollope, W. S. W., & Trollope, L. (2004). Prescribed burning in African grasslands and savannas for wildlife management. *Arid Lands Newsletter*, The University of Arizona. Available at: http://ag.arizona.edu/OALS/ALN/aln55/trollope.html [Accessed 2 May 2024].

Tsuda, S. (2009). Environment of burning grassland – What is 'Yamayaki (grassland burning)'? *Report on Kanpu-zan Symposium*, Tsuda Lab., Gifu University, Available at: http://www.green.gifu-u.ac.jp/~tsu.a/noyaki/KPZ-Archive2.pdf [Accessed 2 May 2024].

Uchino, A. (2016). Plants and animals in Aso. *Survey report of Aso cultural landscape* –Vol II. *Detailed investigations*. Council for Aso Area World Culture Heritage (ed.), Aso: Council for Aso Area World Culture Heritage, 50–74. (In Japanese.)

Yamanoi, T. (1996). Geological investigation on the origin of black soil, distributed in Japan. *The Journal of the Geological Society of Japan*, *102*, 526–544. (In Japanese with English summary.)

Yokogawa, H., & Takahashi, Y. (ed.) (2017). *Agricultural landscape around Mt. Aso and its ecosystem service – From landscape valuation by means of the cultural landscape concept through landscape rehabilitation to World Culture Heritage*. Tokyo: Norintokeishuppan, 1–378. (In Japanese.)

Part 3

Applications

7

CASE STUDY

Rehabilitating the 19th-Century Housing Blocks of Alexandria

Ahmed K. Taher

Bridging the Past and Present: A Heritage Retrofit Strategy for Improved Indoor Performance

Historic buildings, while often possessing inherent design features that promote natural ventilation and passive cooling, frequently struggle to meet the demands of modern occupants for energy efficiency and thermal comfort. This chapter explores this paradox through a case study of a 19th-century residential building in Alexandria, Egypt. Despite its Mediterranean climate and architectural features conducive to natural ventilation, the building heavily relies on air conditioning. By examining this discrepancy, the chapter seeks to uncover the potential of harnessing the building's inherent design and embodied carbon for improved energy performance and occupant wellbeing.

Ventilation methods in buildings play a crucial role in enhancing indoor air quality and thermal comfort, especially during summer months. Adequate air movement provides sufficient air velocity to maintain acceptable levels of thermal comfort in environments with high temperatures and humidity. This chapter will delve into the specific ventilation strategies that can be applied to heritage buildings to optimize their performance while preserving their historical integrity (Santamouris & Asimakopoulos, 1996).

Computational fluid dynamics (CFD) software is employed to simulate and analyse the impact of various retrofit measures on the building's airflow patterns and energy performance. This approach allows for a detailed and quantitative assessment of different options, ensuring that the selected measures not only improve indoor conditions but also align with the building's heritage characteristics.

DOI: 10.4324/9781003527404-10

Alexandria's Urban Fabric and Heritage Challenges

Alexandria's urban character is a product of its rich and layered history, resulting in a complex and multifaceted cityscape. The city center, in particular, is a captivating blend of architectural styles and periods, reflecting Alexandria's evolution to a cosmopolitan hub. This intricate urban fabric, imbued with historical significance and aesthetic value, forms the heart of the city (Awad, 1990; Reid, 2003).

However, the city's heritage is under considerable pressure. Since the mid-20th century, Alexandria has witnessed a rapid transformation marked by uncontrolled development, disregard for historic preservation, and neglect of existing buildings. These challenges have eroded the city's unique character and threatened its cultural identity. Rural immigration, rather than population growth, was a primary factor in the housing shortage that followed the 1952 revolution. As people sought better economic opportunities in Alexandria, many single-family houses were divided into multiple apartments on the same floor. While this provided more housing options, it often compromised the original architectural integrity of the buildings. Despite these adverse conditions, the inherent qualities of Alexandria's built environment offer opportunities for sustainable solutions.

The latest heritage catalogue of Alexandria listed 1,135 conservation buildings that were divided across Alexandria's districts (Alex-Med, 2008; NOUH, 2010). Within the particular typology of buildings defining the above open spaces, an average of 80% are historic with valuable architectural styles and richly detailed façades. Most of the listed buildings were concentrated in Downtown Alexandria around the eastern harbour, Figure 7.1, represented in the Eastern and Central districts, which is a true reflection of the city's historic and urban fabric evolution. Downtown Alexandria acts as the heritage node of the city emphasizing its urban significance.

The city's architectural character was profoundly shaped by eclectic revivalist styles during the late 19th and early 20th centuries, reflecting a cosmopolitan society embracing European influences. Neo-Renaissance dominated, but the city's architectural palette was diverse, including neo-classical and neo-Romanesque elements (Awad, 2010; Yehia, 2006). From this eclecticism, neo-classic revival styles were the dominant styles within the heritage catalogue. Buildings are generally in medium to excellent condition, yet their finishing materials are increasingly deteriorating and require intervention. The current Alexandrian heritage listing is considered a logical reflection of the city's urban architecture both in architectural style and demographic distribution developed throughout the last 150 years, Figure 7.2 (Pallini, 2006; Turchiarulo, 2009).

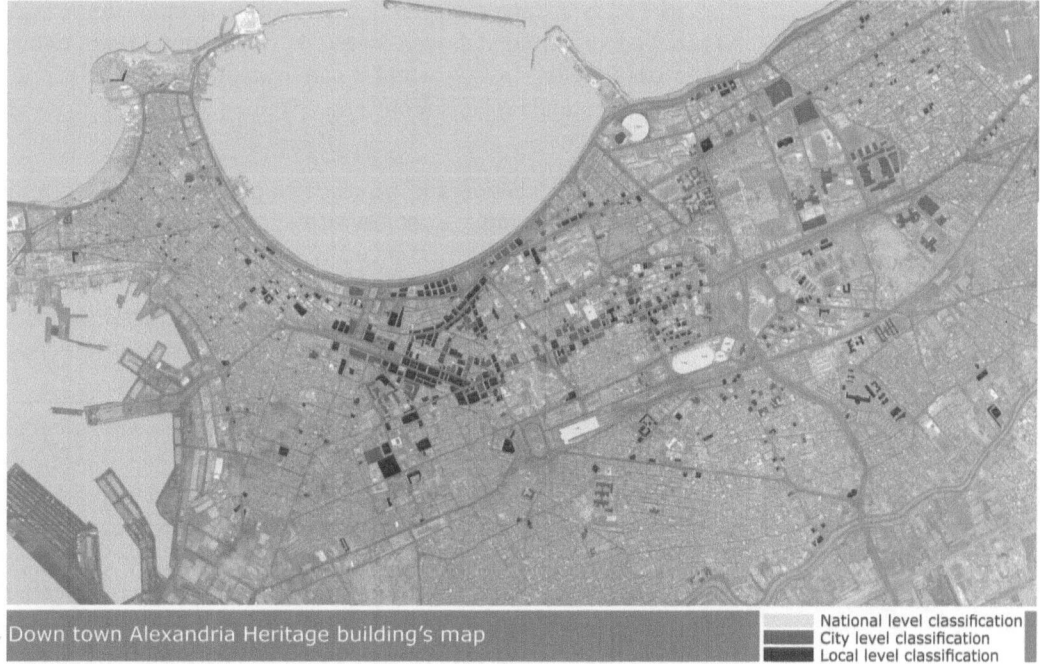

Figure 7.1 Downtown Alexandria heritage buildings map edited by author.
(Alex-Med, 2008)

Figure 7.2 Historic photographs of Place de Consuls by Francesco Mancini.
(Haag, 2008)

Alexandria's Climatic Conditions and Urban Fabric

The city's Mediterranean climate and the architectural characteristics of its historic buildings present a favourable context for exploring natural ventilation and thermal comfort strategies. Alexandria's urban form, characterized by a grid-like layout and a predominance of urban canyons, presents a unique context for exploring natural ventilation.

The city's historical planning, dating back to its foundation in 320 BCE, has shaped its orthogonal street network, which continues to define the core of the city. This consistent urban structure provides a foundation for understanding airflow patterns and their potential impact on building performance.

The city's coastal location and exposure to prevailing northwesterly winds create favourable conditions for natural ventilation. The Mediterranean climate, characterized by hot and humid summers, further emphasizes the need for effective cooling strategies. Alexandria's urban fabric, with its combination of building heights, street widths, and orientation, can influence wind speeds, air pressure, and airflow patterns within the city (Climate, 2019; Shalaby et al., 2017).

Alexandria's distinctive linear morphology, situated within a heritage-rich context, significantly influences the city's relationship to wind. The urban fabric can be broadly categorized into two primary zones: The waterfront and the inland areas. The waterfront zone is defined by a continuous built form that shapes the city's maritime façade. Extending approximately 400 m inland, this zone stretches along the entire coastline within the urban boundaries. In contrast, the inland fabric is characterized by a grid-like pattern of urban canyons oriented at a 20° angle, aligning with the prevailing northwesterly winds. While the waterfront is predominantly built-up, the inland area also features sporadic freestanding structures. This specific urban configuration, with narrow streets perpendicular to the sea and wider canyons parallel to the coast, facilitates wind penetration deep into the city's fabric. This interplay between the built environment and wind patterns is a crucial factor to consider in understanding Alexandria's climate and microclimate conditions, see Figure 7.3.

Figure 7.3 Aerial view and plans highlighting different urban grid orientations with the heritage fabric of the city in relation to prevailing wind direction.
(Ahmed K. Taher)

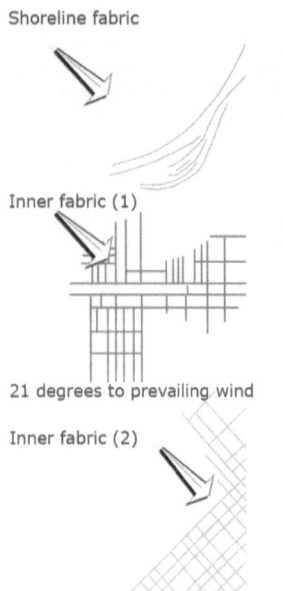

Shoreline fabric

Inner fabric (1)

21 degrees to prevailing wind

Inner fabric (2)

parallel to prevailing wind

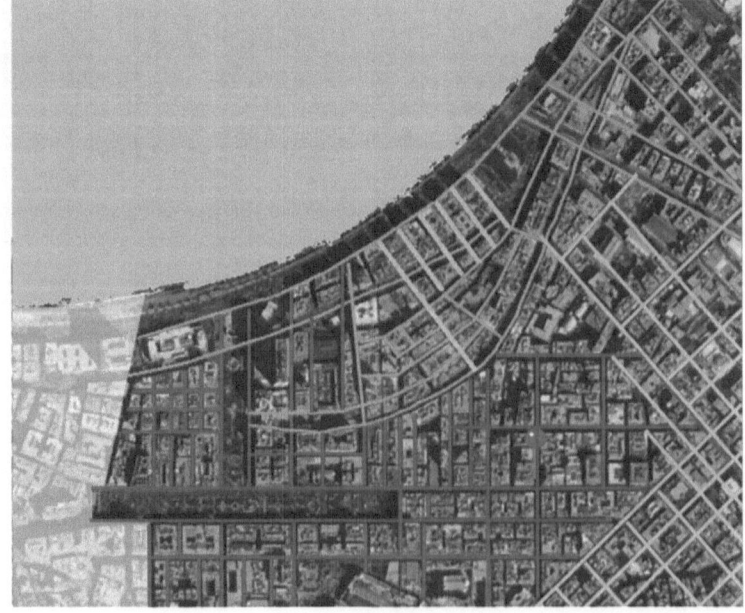

The prevailing northwesterly winds, often reaching speeds of 3.4 to 5 m/s, are channelled through the city's perpendicular street grid, creating stronger wind velocities within the urban fabric. Field measurements have confirmed these patterns, with average wind speeds of 2.4 m/s and higher velocities observed along the coastline. The resulting vortex formation within street canyons enhances ventilation (Weather and Climate, 2019). The hot, humid climate, has temperatures ranging from 28.5°C to 32°C during summer months. Despite high humidity levels, the city's relatively moderate temperatures support the potential for natural ventilation. While the urban environment offers opportunities for ventilation, challenges such as air pollution, noise, and cultural factors must be considered. Despite these constraints, Alexandria's urban fabric, characterized by a chessboard layout, has demonstrated a capacity to mitigate pollution.

In conclusion, Alexandria's climatic conditions, combined with its distinctive heritage urban structure, create a favourable environment for natural ventilation. Understanding these factors is essential for developing effective sustainable strategies for the city's built environment and for resisting the impetus to install carbon intensive systems.

The Selected Case Study Background and Analysis

The building selected for the study is a listed building which is sought to be a representative sample for the heritage buildings in Alexandria built during the same period and to the same architectural style. The building was designed and built with traditional building materials and technologies as a residential four-storey building located within the European district of Alexandria The suggested criteria for choosing the case study building for the research are demonstrated in Table 7.1, according to the building's background, cultural values, heritage listing, and physical accessibility.

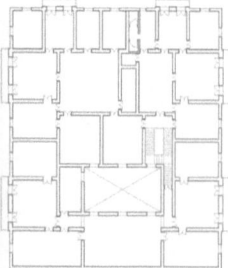

Figure 7.4 Selected case study building, building image, plan, and façade.
(Ahmed K. Taher)

The choice of this building as a case study, which is representative of the majority of the current listed buildings within the city, as a local level classification residential block (around 75% of the listings), was intentional. As a consequence, the methodology could be applied to

Table 7.1 Description of the case study heritage building

Criteria	Description		
Background	Date	1869	
	Original owner	Princess Najwan, currently owned by the government	
	Current use	Residential, ground floor commercial	
Cultural values	Social evidential	A representative place of social life of upper-class families who lived in Alexandria. It was an upper-class residential building	
	Architectural	Value	Represents the multi-cultural value of the cosmopolitan Alexandria heritage context
		Style	The neo-classic eclectic revivalist architecture style, the dominant style within the context
Heritage listing	Listing number	Listed number according to the catalogue of 2007 by Alexandria Preservation Trust '108'	
	Listing grade	Grade (C) local level classification, representing 75% of the city's heritage	
	Location	The case study building lies within the heart of the heritage context of the city, located on Sizostriss and Msjid el Attarin, both are listed as conservation streets	

(Ahmed K. Taher)

the wider local context as it is considered to share the same typology and level of intervention. Moreover, results of the research will be valid within a certain level on the wider scale within the cosmopolitan heritage fabric of the city.

Heritage-Sensitive Natural Ventilation Retrofit

To enhance the natural ventilation performance of the selected building while preserving its heritage value, we adopted a strategic approach. This strategy was informed by the building's cultural significance within Alexandria, its heritage value, and a comprehensive assessment of its current natural ventilation performance.

- Understanding the building's original layout and ventilation systems.
- Evaluating the building's current ventilation performance.
- Historic reconstruction of the original building performance.
- Exploring various passive natural ventilation systems, and assessing their suitability based on their impact on heritage value and ventilation performance.

By following this strategic approach, we aim to develop a tailored retrofit plan that effectively improves the building's natural ventilation performance while safeguarding its heritage.

Analysing Potential of Natural Ventilation in a Heritage Alexandrian Building: A Case Study

The Building Current Layout Internal Space Categorization

The building's design represents a stark contrast to the contemporary, air-conditioned, multi-storey structures prevalent in Alexandria. Its inherent passive design features, such as large, well-ordered openings, an atrium, and a staircase, suggest a focus on cross-ventilation and natural cooling.

The building's dimensions are 34 × 30 × 22 m, with a footprint of 1020 square meters. It features a light well measuring 10.5 × 6 × 22 m. The building envelope is characterized by varying window sizes, with a total opening percentage ranging from 10 to 17%. Windows are centrally located within indoor spaces. While internal doors have high-level openings, site surveys revealed that they are often blocked by occupants.

The building's typical floor plan comprises three flats. To analyse the airflow dynamics within these flats, the internal spaces were categorized into depth zones based on their proximity to external openings. Zone S1 has direct openings to the exterior, while zone S2 has no direct openings, relying on internal shafts. Interestingly, the S2 zone, despite its limited external exposure, is currently used for primary living spaces (Figures 7.5 and 7.6). In addition to categorizing the depth map, auditing points were placed in different zones. The points were used for CFD monitoring for specifying the airflow speed within each zone to achieve a complete survey of the detailed floor plan internal spaces Figure 7.7.

Figure 7.5 Photographs showing inner openings.
(Ahmed K. Taher)

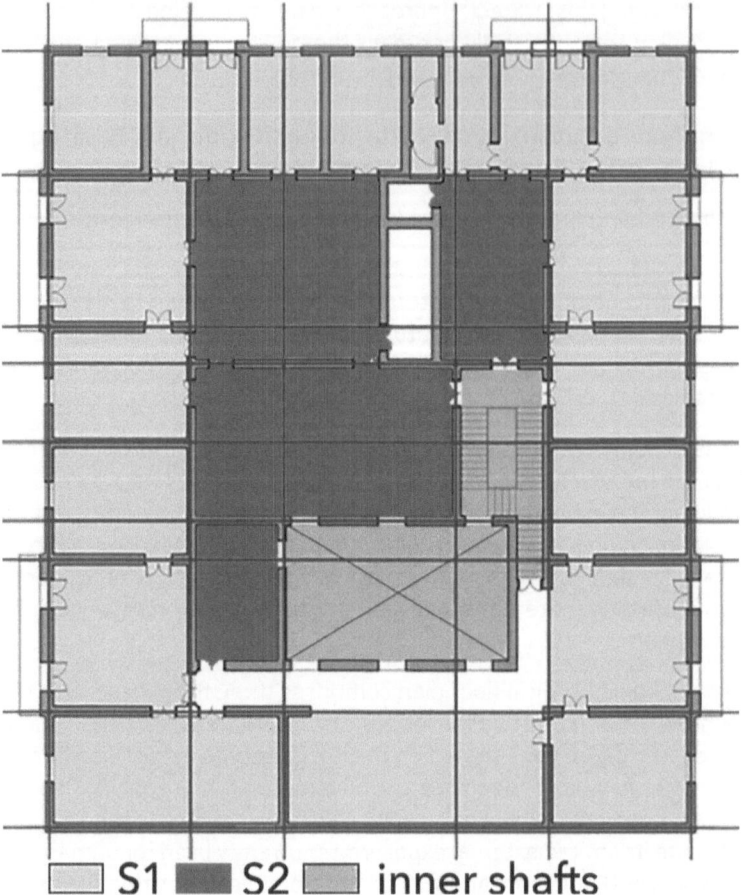

☐ S1 ■ S2 ☐ inner shafts

Figure 7.6 Plan showing different spaces' opening configuration.
(Ahmed K. Taher)

Building Mass, Wind Orientation, and Canyon Configuration

The block is a compact deep mass with dimensions of L = 34 m, W = 30 m and H = 22 m, and a W/L 0.882. The block consists of four storeys high with a footprint area of 1,020 m² and volume of 22,440 m³. The block contains a courtyard penetrating the mass with 57.75 m². This court-yard is used to provide light and ventilation to the inner spaces. The nature of this deep plan design of the block is against the recommen-dation derived from the literature and may reduce the potential of using ventilation. Wind angle of attack is south west, with an angle of 112° clockwise. Although the prevailing wind in the Alexandrian con-text is directed from the north-west. The urban canyon configuration for Sizostris Street and Masjid el Atarin Street are H/W 1.96 and 1.78, respectively, having a deep configuration on both streets, enhancing the potential for natural ventilation – see Figures 7.8, 7.9, and 7.10.

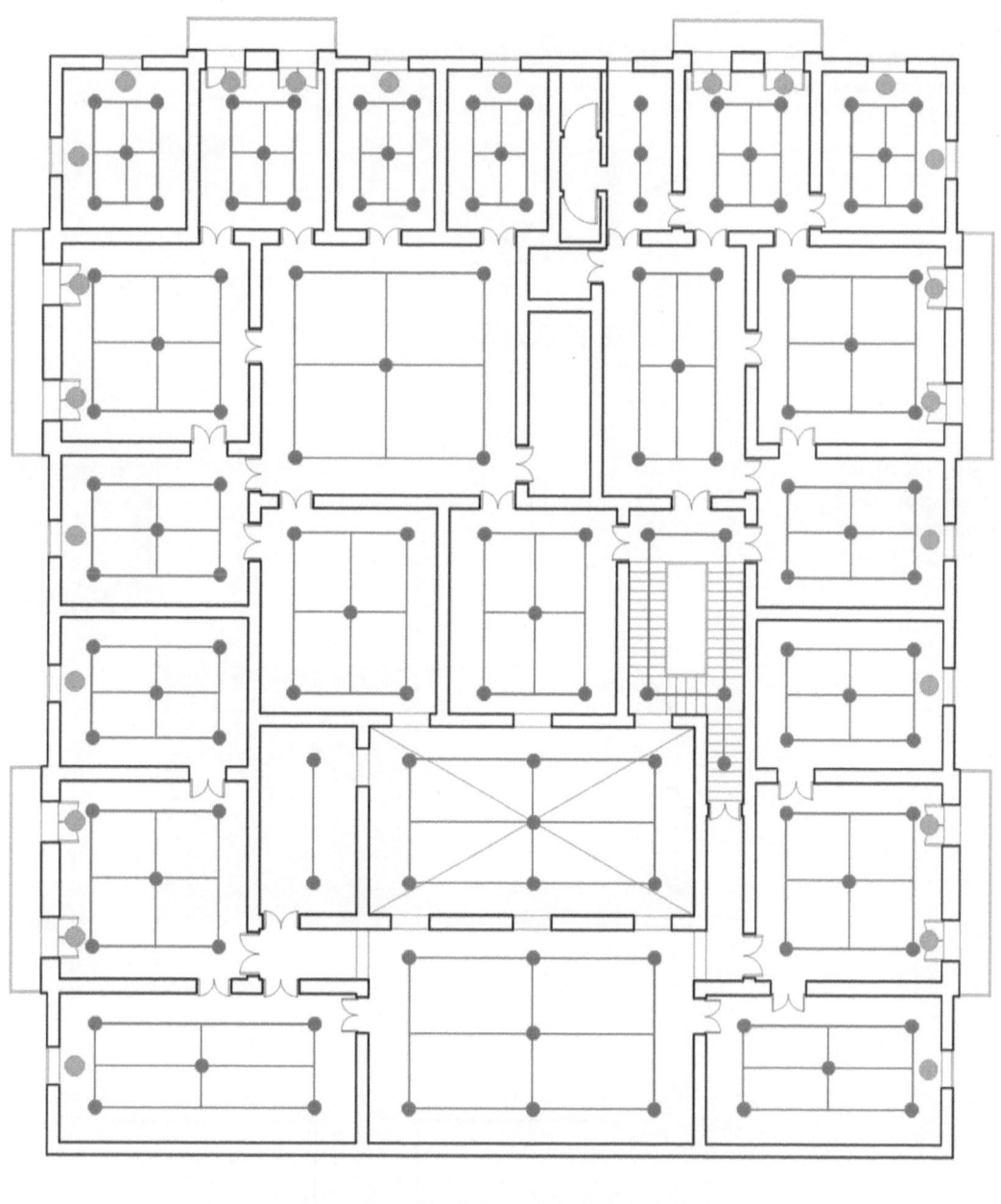

● external openings air apeed monitor points   ● internal air speed monitor points

Figure 7.7 Detailed floor plan highlighting the monitoring points.
(Ahmed K. Taher)

**Figure 7.8 Plan and
isometric of case study
building form and mass.**
(Ahmed K. Taher)

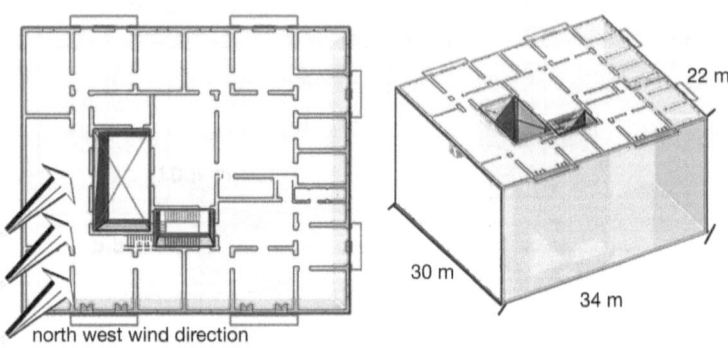

north west wind direction

22 m

30 m

34 m

**Figure 7.9 Plan
showing prevailing
wind direction.**
(Ahmed K. Taher)

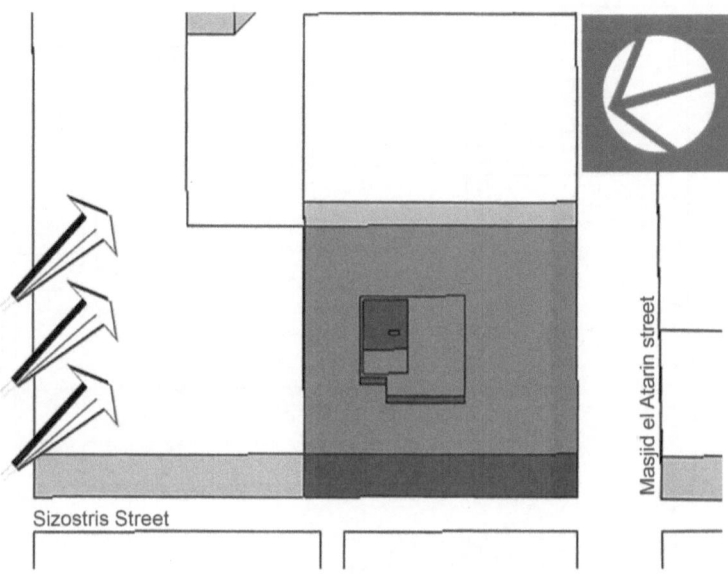

Masjid el Atarin street

Sizostris Street

wind to street relation 22-112 degree relation

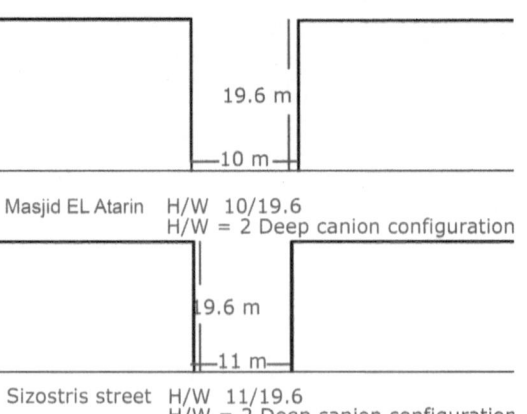

19.6 m

10 m

Masjid EL Atarin   H/W 10/19.6
                    H/W = 2 Deep canion configuration

19.6 m

11 m

Sizostris street   H/W 11/19.6
                    H/W = 2 Deep canion configuration

**Figure 7.10 Plan
showing urban canyon
configuration.**
(Ahmed K. Taher)

Building Envelope and Internal Openings

The case study building has a flat roof with no vertical projection designed for ventilation. There is a courtyard to drive air pressure and induce the air into internal spaces. In terms of openings design, the size, position, and morphology of the openings were analysed. Two different sizes are identified with inlet percentage in relation to the inner spaces, ranges from 5 to 36%. In terms of position, the windows are located in the centre of the space, there are two balcony openings which are also located in the centre of the space. Morphologically, as case study openings are of the vertical-vane opening type with double-side hinges all windows and balcony openings have external venetian shutters – see Figure 7.11. In addition, all internal doors are solid timber frame with a high-level transom opening which were modified and closed by occupants.

Figure 7.11 Plan and elevation showing opening position arrangement and opening percentage in relation to the space. (Ahmed K. Taher)

Assessment Methodology and Current Performance Evaluation

The assessment of the case study building involved two primary stages: A detailed monitoring phase and a computational fluid dynamics (CFD) analysis. (A) Monitoring and Validation, the initial stage focused on collecting detailed field measurements to validate the CFD model's input parameters. Airflow data was gathered both inside and outside the building. By comparing these measurements with the CFD simulation results, the accuracy and reliability of the numerical model was evaluated. This validation step ensured that the CFD parameters were suitable for subsequent, more complex simulations. (B) CFD Analysis, the second stage employed CFD software, ANSYS Fluent, to investigate airflow patterns in and around the building. A simplified volumetric representation of the building was created for the simulation, which incorporated both the external environment and the internal spaces. The CFD model was validated against field measurements, focusing on air speed data. By analysing the airflow magnitude and patterns within different internal spaces, the building's current performance deficiencies and potential for improvement were identified.

Field Monitoring

To measure wind speeds, two hotwire anemometers were deployed at specific locations within and around the case study building. These devices, manufactured by Airflow model TA 2, have a stated accuracy of ±3%. Due to limited availability, only spaces with direct external exposure (S1) were monitored using points P1 and P2. Field measurements were conducted at various locations on the building's northwest and south-west façades, including both the first floor and the rooftop. The anemometers were positioned as indicated in Figures 7.12 and 7.13. To calculate U and Uref speeds, pairs of anemometers were placed at different heights: P1 and RP1 were located 7.5 m above ground level inside the building and 22 m above ground level on the

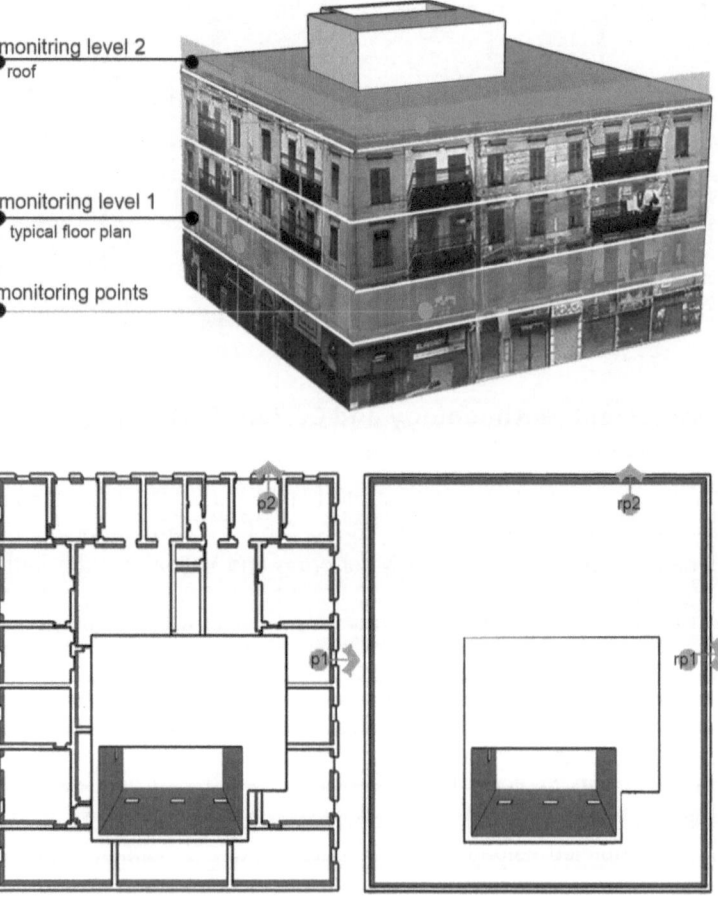

Figure 7.12 Isometric and plans showing different monitoring levels and points.
(Ahmed K. Taher)

Figure 7.13 Photographs showing different monitoring levels and points.
(Ahmed K. Taher)

roof, respectively. Similarly, P2 and RP2 were positioned on the south-west side of the building. The measurements were conducted with all external openings fully open to assess the maximum potential for natural ventilation. This setup was later replicated in the CFD simulation to validate the field measurement results.

The field measurements were conducted during the month of July to capture summer conditions. To ensure data reliability, airflow velocity was measured for a consecutive week, with readings taken at specific times (4:00 PM and 4:30 PM) for each pair of anemometers (P1, RP1 and P2, RP2). Each measurement session lasted 30 minutes, and velocity magnitudes were recorded at 30-second intervals, aligning with literature recommendations for accurate averaging.

CFD Simulation

For the CFD simulation, Ansys Fluent 18.1 software was used to perform the computations for the assessment and evaluation of the air movement patterns in and around the case study building. Although it is recognized that CFD results are subject to uncertainties and approximations, the achievement of consistency is related to the control of a number of input parameters.

Model and Domain

A model for the case study building and the surrounding building blocks were constructed. All surrounding blocks were modelled as solid blocks except for the monitored building which was detailed. Figure 7.14 shows that only floor two of the building was modelled including the openings and interior partitions/opening. All other floors of the case study building were modelled as solid blocks. The dimensions of the solution domain was set according to Blocken, 2015. Dimensions of H inlet direction: 5 H from the outflow direction and height, where H is the model height (642 × 532 × 100 m).

$$U = Umet. K. z^a \tag{1}$$

In which Umet is the velocity of wind from the meteorological data, K and a are the coefficients of the terrain. The variables K and a were assigned the values of 0.21 and 0.68 for the dense urban site. Free wind at height 30 m was set to 2.05 m/s after correction flowing from the north south direction (incident angle = 22°).

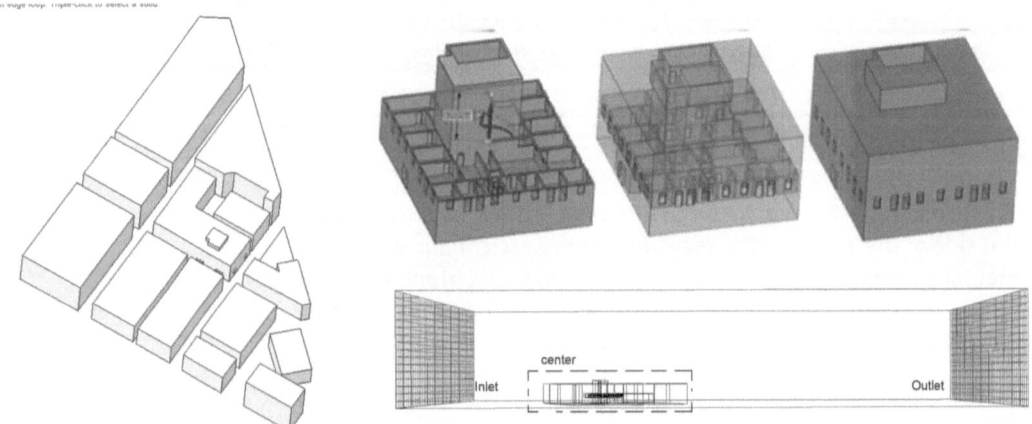

Figure 7.14 CFD modelling and setting the boundary conditions domain.
(Ahmed K. Taher)

Mesh Structure and Solving Parameters

In order to achieve reliable results in the CFD, an initial group of CFD simulations were carried out and results were compared to field measurements collected at locations P1, P2, and P3. The aim was to assess the impact of mesh structure on the contour plot output and either tetrahedral or hexagonal meshes, coarseness/refinement level influence on the results were utilized and all the CFD domains have been designed for minimum blockage, with an average value of 3.0% and a maximum of 4.6% (see Figure 7.15). Initial mesh (A) is too coarse and results are not accurate on both grid options (hexagonal, tetrahedral grids). After the first adaptation (B) the uneven cell distribution in the tetrahedral mesh solution becomes more apparent, impacting on the pressure distribution and later mesh refinement (C) doesn't significantly improve the results, though it increased simulation computer time. A sequence of two refinements proved to be enough to allow results to become independent of mesh size. The solution solver was set as pressure based, and was of implicit mode for steady-time problems. The turbulent viscosity model adopted for all the CFD simulations was K-ε RANS standard, in which average speed and turbulent intensity profiles were used

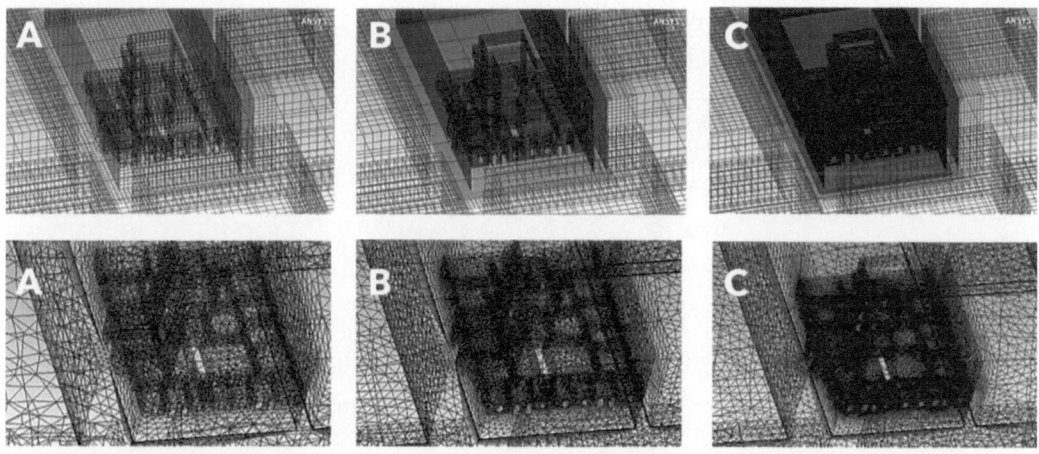

Figure 7.15 Mesh structure for CFD modelling: Different monitoring levels and points; different mesh refinement for hexagonal (above) and tetrahedral (below) mesh type applied.
(Ahmed K. Taher)

Setting the Boundary Conditions

After building the case study's model, the boundary conditions of the model had to be specified. These conditions include solution domain configurations, ambient conditions, the system's fluid properties, turbulence model, solution type, and wind boundary profile. The following ambient conditions were attached to the solution domain:

- The turbulent model used is standard 3D RANS model with standard wall treatment
- The gravitational in the normal value of (9.81 m/s^2) and the normal direction (–Y)
- Ambient temperature of 32.5°C in July (according to the nearest weather station)
- The fluid was set for air at constant density (1.19 kg/m^3) and viscosity (1.79e-0.5kg/m-s).
- The operating pressure conditions of the domain were kept at 101325Pa.
- The dimensions of the solution domain (642 m, 532 m, 100 m).

The boundary types used are velocity inlet, interface, non-slip walls, and outflow boundaries (Franke et al., 2010). The upstream boundary was set at 'velocity inlet', an ABL profile was imposed with the ABL calculated by the logarithmic profile for a given terrain roughness according to Reynolds numbers. Free wind after correction flowing from the north west had an incident angle of 22°. After setting the boundary conditions, the problem was initialized, the monitoring plot was set for monitoring four critical points in the model in addition the pressure coefficient on the case study building to ensure the reliability of the solution.

Figure 7.16 Simulation convergence plot.
(Ahmed K. Taher)

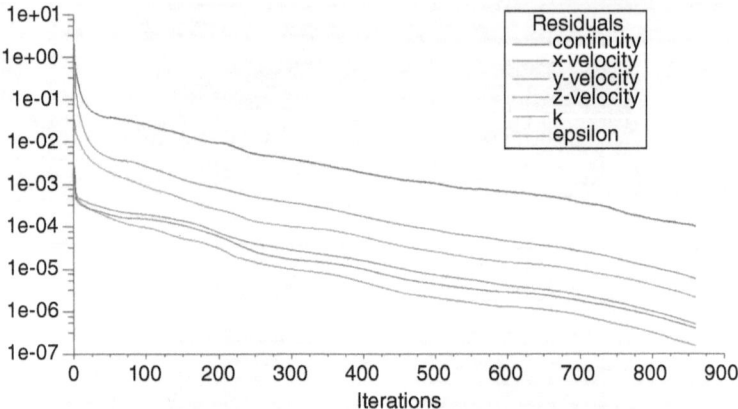

The results revealed that a value of four orders of magnitude convergence was obtained, the steady state solution was converged in 7 hours and 47 minutes with 872 iterations see Figure 7.16. The visual and the numerical results of the airflow patterns and airspeed in and around the case study building were then extracted and analysed.

Results from the CFD simulation were cross compared to the actual monitoring data gathered from P1, P2, and the roof points (1) and (2) monitors (Table 7.2), while reviewing the results of specified points on both simulation and monitoring with a factor error of 6%, an acceptable error for the simulation.

The airflow inside the first floor of the monitored block was further analysed using the outcome from the CFD simulation. The primary goal is to evaluate the impact of building configuration, external openings, and occupant behaviour on indoor air quality. The building's external openings are not directly exposed to prevailing winds. The north-west and south-east elevations act as positive pressure openings, with the north-west elevation allowing higher wind velocities. The south-west elevation acts as a negative pressure opening. The simulation considers the building's detailed plan and occupant behaviour regarding opening and closing of external openings and doors. The simulation is conducted with all external openings open and internal doors simulated based on occupant privacy.

Table 7.2 Monotiling of points velocity acquired from the CFD model

Monitoring point	Physical monitor: Airflow velocity m/s	Simulation RANS model's airflow results m/s
		K-ε standard model
Roof point (1)	1.71	1.68
Roof point (2)	1.12	1.15
P1	0.69	0.72
P2	0.59	0.57

(Ahmed K. Taher)

The airflow within the typical floor is generally poor, with significant separation between different zones (S1 spaces, S2 spaces, inner court-yard, and staircase) see Figures 7.17, 7.18, 7.19, 7.20, and 7.21. The main air sources are the S1 spaces on the outer boundary with external openings. There is no cross-ventilation between external openings and the inner courtyard due to the lack of connections between different zones. S1 spaces perform better due to their direct relation with the external environment. S2 spaces, the main living spaces, are poorly ventilated due to their complete separation from the outer environment and inner courtyard. Airflow within S1 spaces often forms vortices and exits through the same side as it enters. Airflow within S2 spaces is nearly stagnant, with low air speeds (0.04–0.21 m/s) see Table 7.3.

Table 7.3 Internal airspeed inside the detailed current performance floor plan spaces

Zone		Inlet airspeed (m/s)	Internal airspeed (m/s)		Average airspeed (m/s)
			max	min	
S1	1	2.1	1.35	0.77	1.41
	2	0.13	0.19	0.07	0.13
	3	0.32	0.19	0.13	0.21
	4	0.28	0.25	0.08	0.21
	5	0.37	0.18	0.09	0.21
	6	0.12	0.13	0.07	0.11
	7	0.39	0.33	0.12	0.28
	8	0.42	0.15	0.18	0.25
	9	0.48	0.18	0.05	0.23
	10	0.33	0.19	0.05	0.19
	11	0.32	0.15	0.05	0.17
	12	0.35	0.11	0.08	0.18
	13	0.24	0.19	0.06	0.16
	14	0.24	0.23	0.17	0.21
	15	0.3	0.28	0.12	0.23
	16	0.28	0.23	0.14	0.22
	17	0.12	0.11	0.07	0.11
S2	1	-	0.16	0.11	0.14
	2	-	0.21	0.04	0.13
	3	-	0.07	0.03	0.05
	4	-	0.04	0.03	0.04
	5	-	0.04	0.03	0.04
	6	-	0.16	0.06	0.11
I	1	-	0.84	0.35	0.59
	2	-	0.62	0.18	0.41

(Ahmed K. Taher)

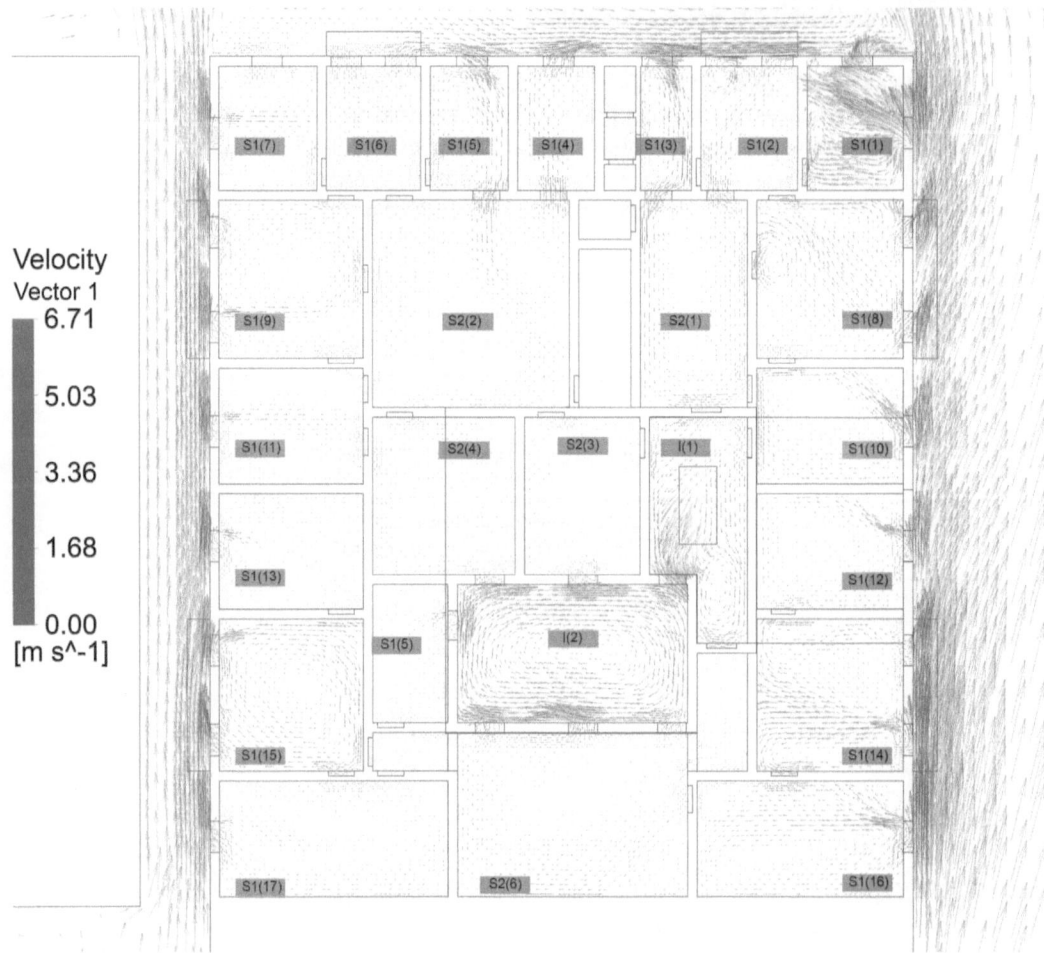

Figure 7.17 CFD simulation showing airflow pattern inside the detailed floor.
(Ahmed K. Taher)

This analysis of natural ventilation performance of the building examines the building's original layout and its potential for comfort ventilation, considering factors like the surrounding environment, inner space height, inner shafts, and opening ratio. The primary goal is to reduce cooling loads and energy consumption during summer. However, the analysis reveals that the building's current layout and occupant behaviour result in unacceptable indoor comfort conditions. This is primarily due to modifications made to the building's original design, such as the blockage of upper openings, which have negatively affected cross-ventilation and the stack effect. The results provide a detailed example of how a deficiency of performance in natural ventilation can be created in a heritage building.

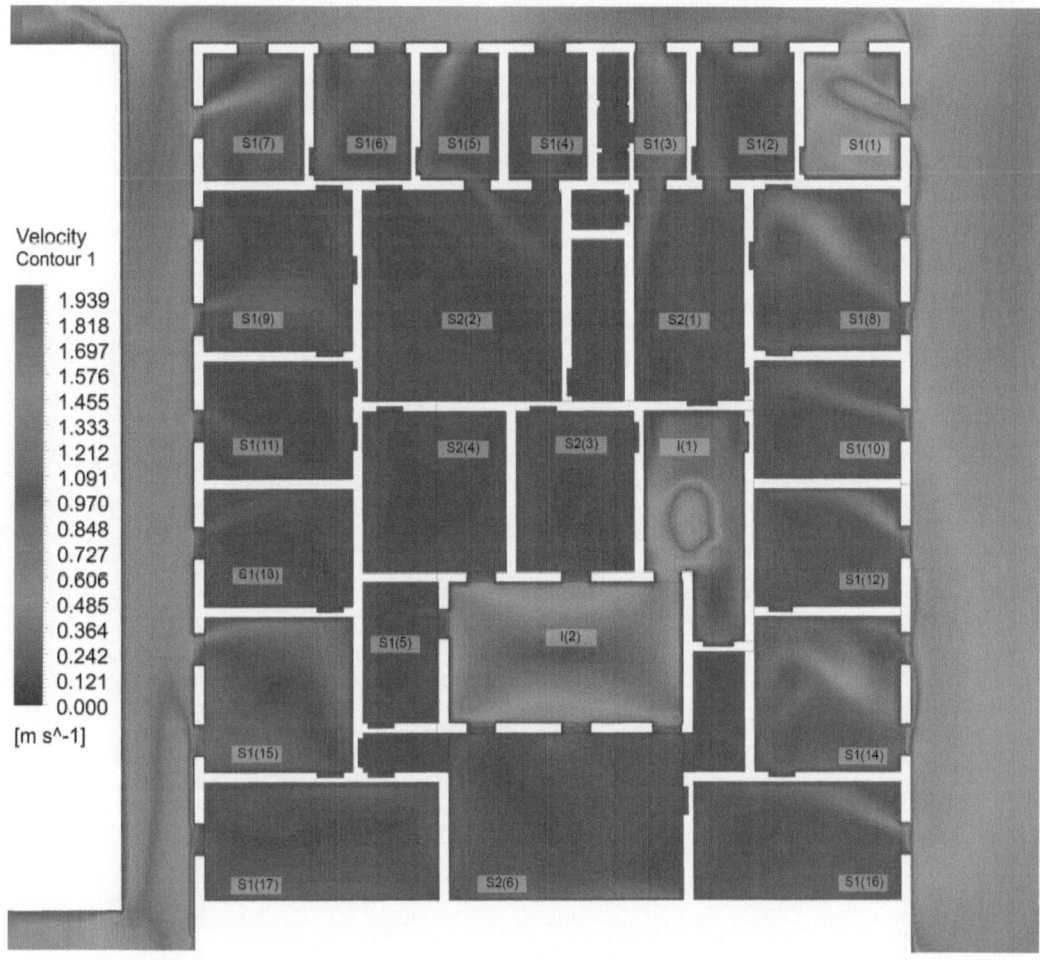

Figure 7.18 CFD simulation showing airflow speed profile of the detailed floor. (Ahmed K. Taher)

Historic Reconstruction

Archival research of the building's history reveals that it was confiscated from its previous owner, Princess Shwikar after the 1952 revolution in accordance with the previously discussed changes to the Egyptian political, economic, and social fabric at the time. The building's occupants were composed of the city elites and foreigners with their migration and the rural relocation toward the city. With the high-density growing population, the building's original layout was altered; this was a common situation within the heritage fabric of the city. The original layout of the building's typical floor was changed from one flat occupied by one family for each floor to be divided into three flats and occupied by three families, the current building's layout and usage. The ground floor usage was the same as the current layout composed of commercial shops. The flat original layout was

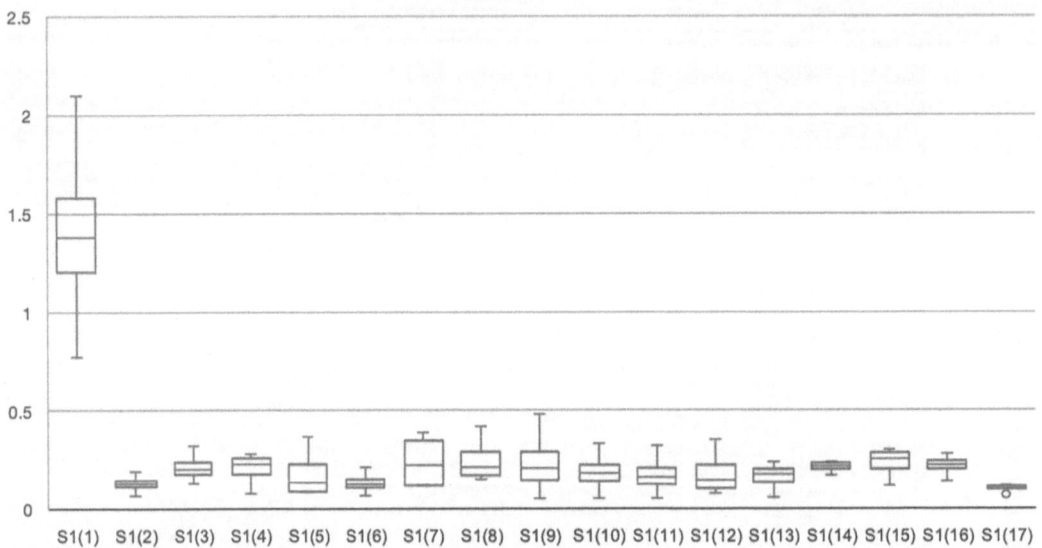

Figure 7.19 Graph showing average velocity m/s for the S1 spaces.
(Ahmed K. Taher)

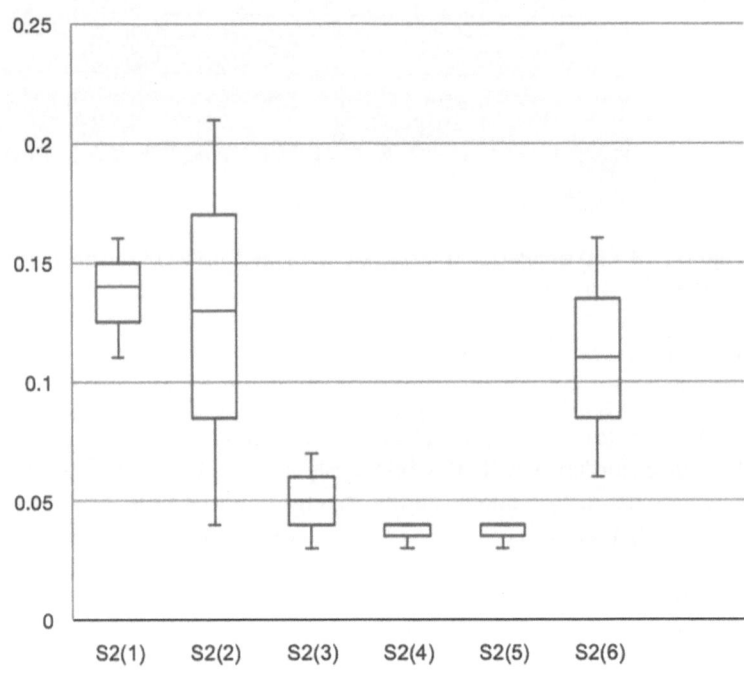

Figure 7.20 Graph showing average velocity m/s for the S2 inner spaces.
(Ahmed K. Taher)

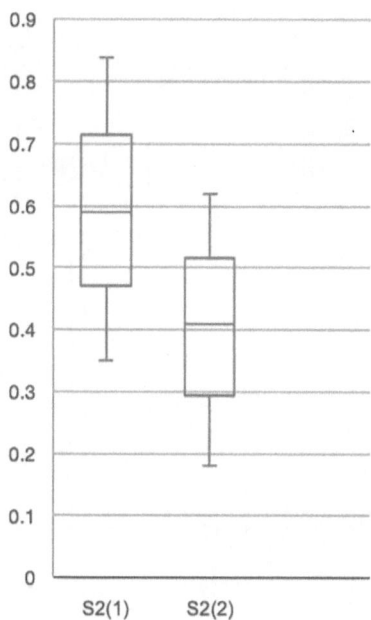

Figure 7.21 Graph showing average velocity m/s for the S2 inner spaces.
(Ahmed K. Taher)

spacious accommodating the city elites, Figure 7.22. The spatial configuration of the flat was composed of a large reception and dining area in the centre, occupants' bedrooms and guest hospitality rooms located on the main streets, while the services including kitchen, storage, and servant's accommodation are found on the back area with a separate entrance. The surrounding context has been slightly changed with the erection of higher new buildings following the recent transformation of the city after 1952.

Simulations of the building's original layout, featuring spacious internal spaces and internal transom windows, demonstrated significantly improved natural ventilation compared to its current configuration see Table 7.4 and Figures 7.23, 7.24, 7.25, 7.26, 7.27. Unlike the current layout, which divides the first floor into multiple, smaller apartments, the original design treated the entire floor as a single unit. This interconnectedness facilitated cross-ventilation between spaces, resulting in enhanced airflow patterns and magnitudes throughout the building. The original layout's larger, open spaces and greater connectivity allowed for more efficient air circulation. This is particularly evident in the increased airflow within the internal spaces, which were now more effectively connected to both external and internal areas. By restoring the existing transom windows and opening specific windows external space S1, additional cross-ventilation patterns were created. This allowed for improved airflow into the internal spaces S2 compared to the stagnant air observed in the current configuration. The external S1 spaces, benefiting from their connection to the internal zones, exhibited higher airflow rates than the current building. This resulted in an increased average internal air speed of 0.57 m/s

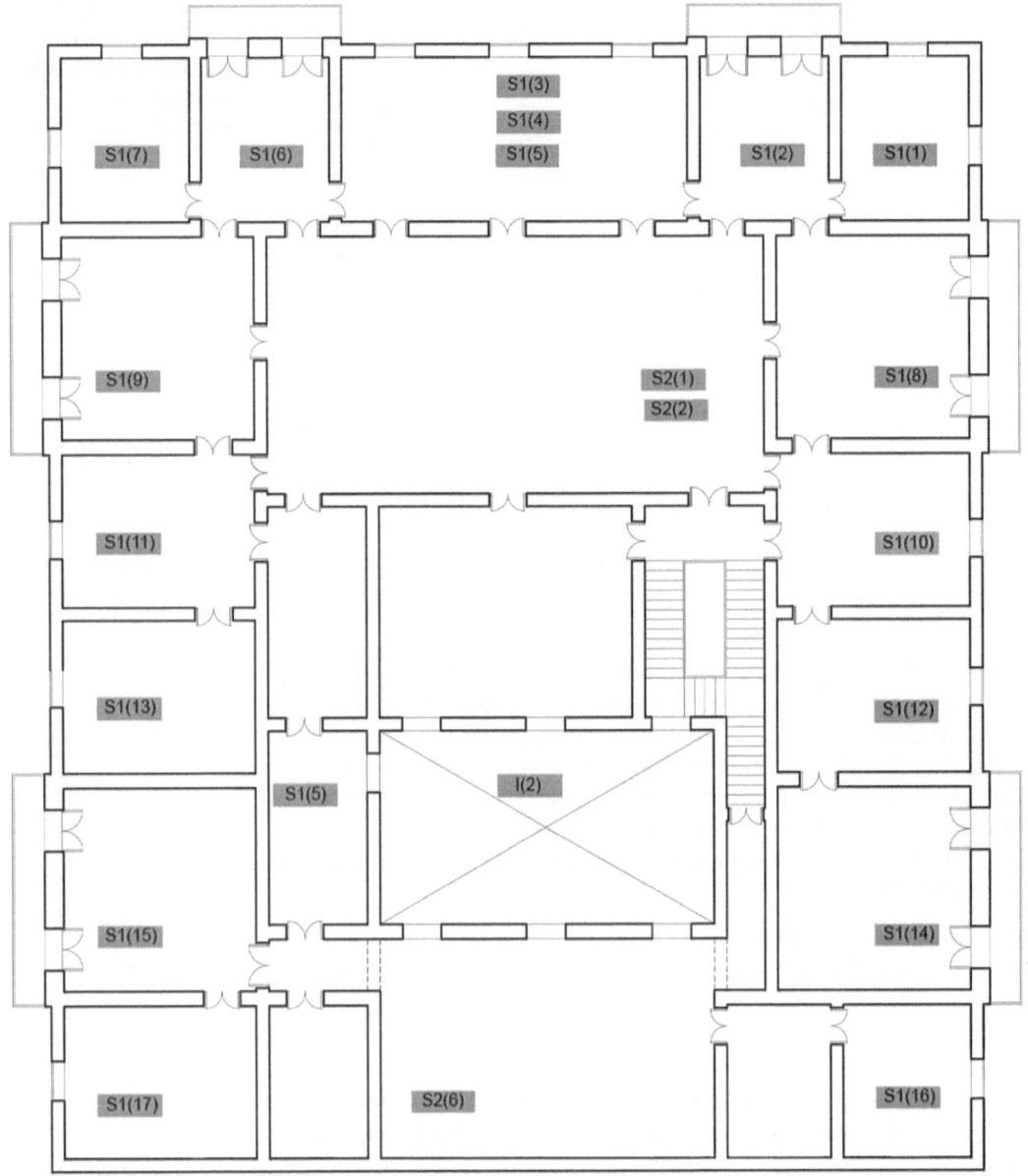

Figure 7.22 Case study building layout before modification.
(Ahmed K. Taher)

compared to 0.26 m/s. Additionally, the internal S2 spaces, which are primarily used for living, experienced improved ventilation from the outer S1 spaces, with a maximum internal air speed reaching 0.44 m/s compared to 0.08 m/s.

When compared to the original layout, the deficiencies in the current building become apparent. The original design's emphasis on connecting internal and external spaces, along with the inner atrium,

Table 7.4 Internal airspeed inside the detailed historic reconstruction of original floor plan spaces

Zone		Inlet airspeed (m/s)	Internal airspeed (m/s)		Average airspeed (m/s)
			max	min	
S1	1	2.2	1.61	0.94	1.58
	2	0.81	0.64	0.44	0.63
	3	0.73	0.47	0.32	0.51
	4				
	5				
	6	0.49	0.66	0.32	0.49
	7	0.53	0.39	0.16	0.36
	8	0.55	0.55	0.36	0.48
	9	0.66	0.26	0.20	0.38
	10	1.97	0.64	0.29	0.96
	11	1.15	0.56	0.27	0.66
	12	0.36	0.22	0.16	0.25
	13	0.81	0.62	0.39	0.61
	14	0.63	0.88	0.42	0.64
	15	0.43	0.39	0.26	0.36
	16	1.12	1.74	0.81	1.22
	17	0.64	0.48	0.35	1.47
S2	1	-	0.92	0.58	0.75
	2				
	3	-	0.57	0.28	0.43
	4				
	5	-	0.31	0.17	0.24
	6	-	0.42	0.27	0.35
I	1	-	0.96	0.46	0.71
	2	-	0.79	0.34	0.57

(Ahmed K. Taher)

created a stable cross-ventilation system. To improve the current building's performance, it is evident that restoring elements of this original concept, such as reconnecting spaces and utilizing the atrium, would be crucial.

Balancing Heritage Preservation and Natural Ventilation: A Retrofit Strategy for the Case Study Building

The case study building's current natural ventilation performance, as assessed through modelling and measurements, is inadequate for

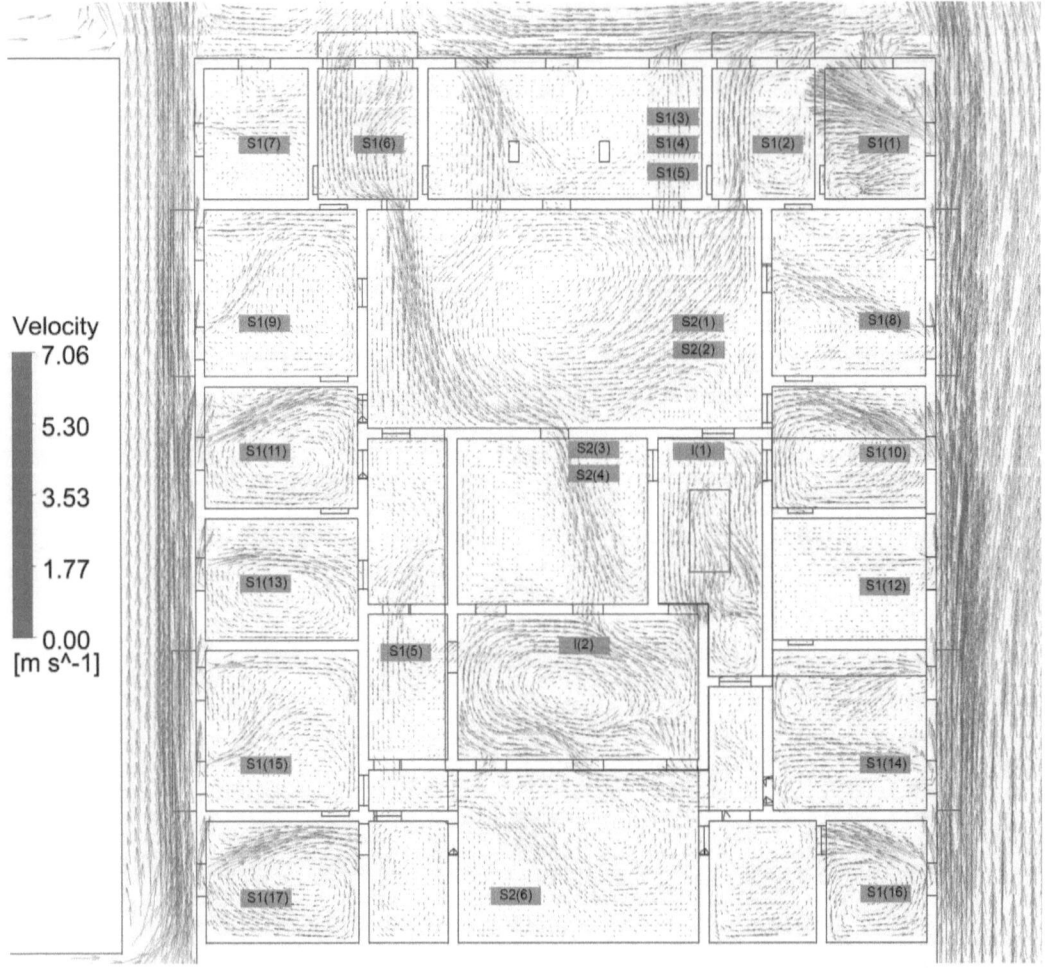

Figure 7.23 CFD simulation showing airflow pattern inside the detailed floor of the original building before modification.
(Ahmed K. Taher)

comfortable indoor conditions. This deficiency is primarily due to the building's inability to induce cross-ventilation and the stack effect, which have been negatively impacted by modifications made by occupants. To improve the building's natural ventilation, passive retrofit strategies are necessary. However, given the building's heritage status, these strategies must be carefully considered to avoid compromising its architectural value, aiming to identify and implement appropriate retrofit measures that enhance natural ventilation while preserving the building's heritage characteristics. This involves balancing the requirements of passive retrofitting with the limitations imposed by the building's heritage classification.

The case study building's heritage classification allows for limited alterations to the external façade but provides more flexibility for

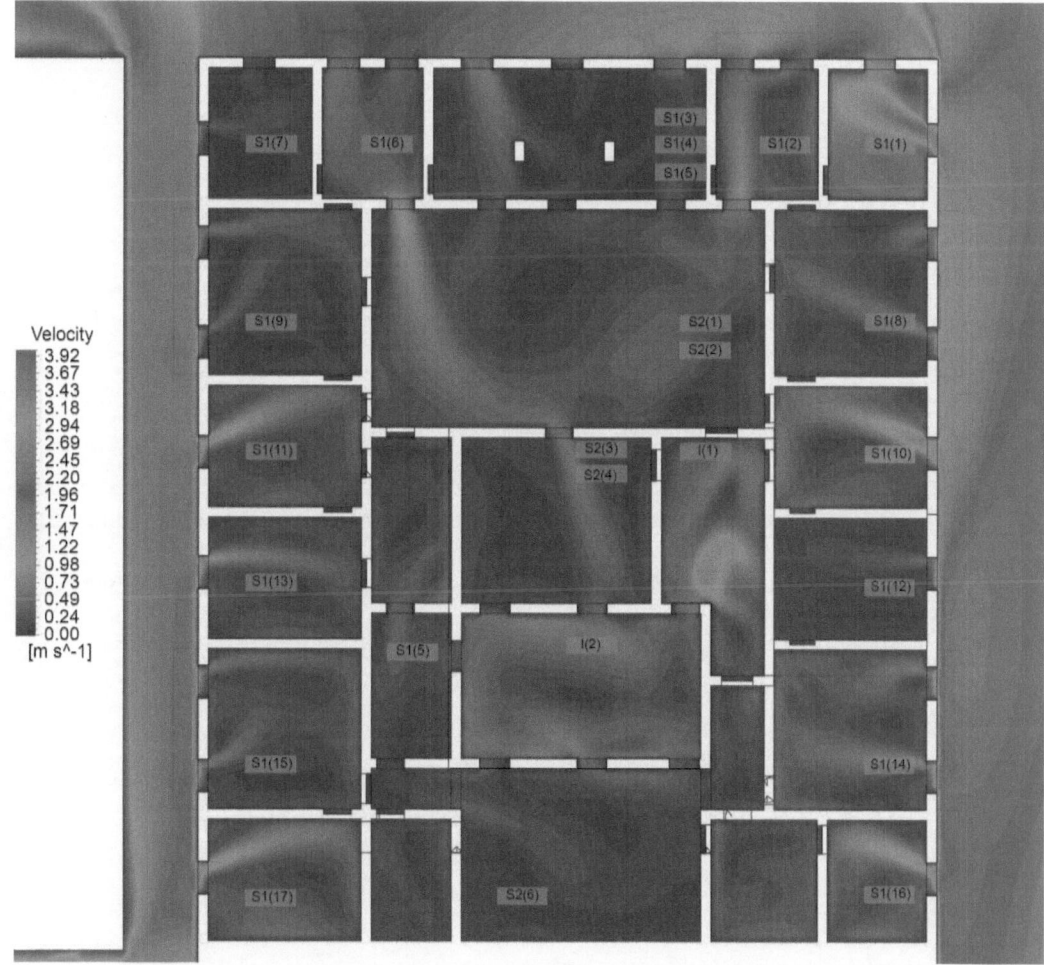

Figure 7.24 CFD simulation showing airflow speed profile of the detailed floor of the original building before modification.
(Ahmed K. Taher)

internal modifications. While the interior is not legally protected, any proposed changes must still be considered in relation to the building's heritage value. focusing on identifying modifications that improve natural ventilation while maintaining the internal spatial configuration and load-bearing wall structure. This process will involve evaluating the building's current performance and comparing it to its architectural identity and heritage value level.

The most obvious observation in the case study building's current ventilation issues is largely attributed to occupant modifications and the division of the floor plan into separate apartments. These changes have hindered cross-ventilation and the stack effect, negatively impacting airflow throughout the building. The internal space categorization analysis revealed that the ventilation of certain spaces (S2) is

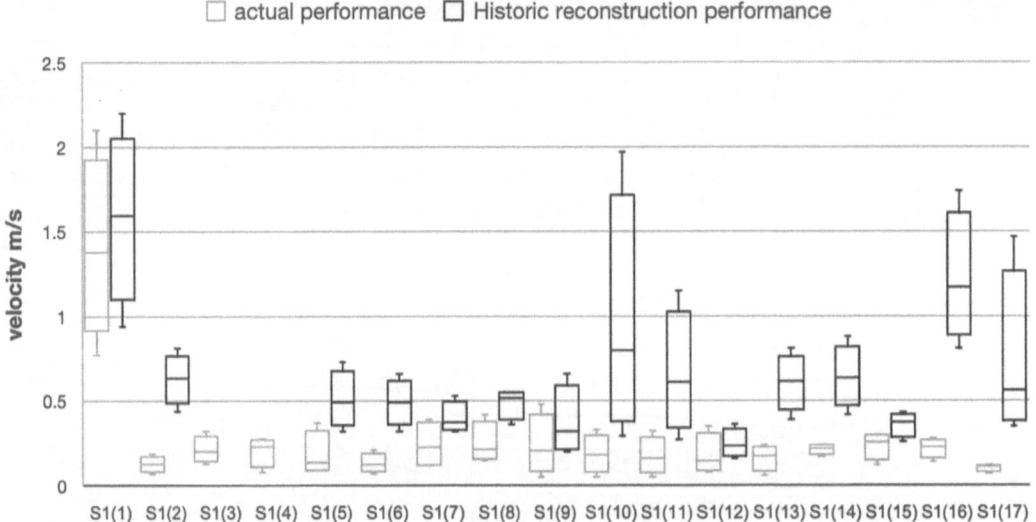

Figure 7.25 Graph showing average velocity m/s for the S1 spaces of the original building before modification.
(Ahmed K. Taher)

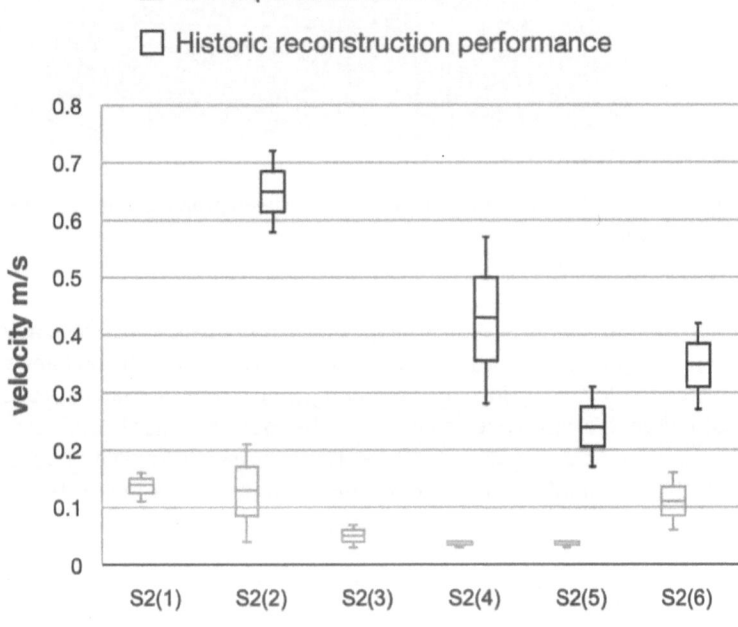

Figure 7.26 Graph showing average velocity m/s for the S2 inner spaces of the original building before modification.
(Ahmed K. Taher)

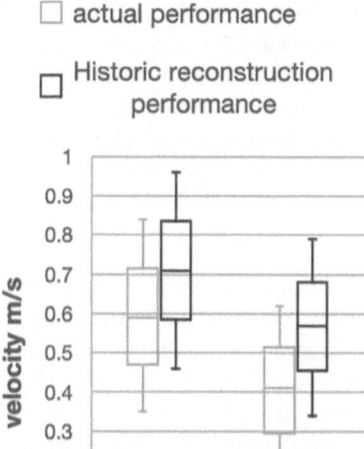

☐ actual performance

☐ Historic reconstruction performance

Figure 7.27 Graph showing average velocity m/s for the S2 inner shafts' spaces of the original building before modification. (Ahmed K. Taher)

dependent on others (S1), as well as the availability of the inner court-yard and staircase. The presence of transom windows in the original design could have facilitated airflow between these spaces. However, the blockage of these windows has limited cross-ventilation and reduced the effectiveness of stack effect ventilation.

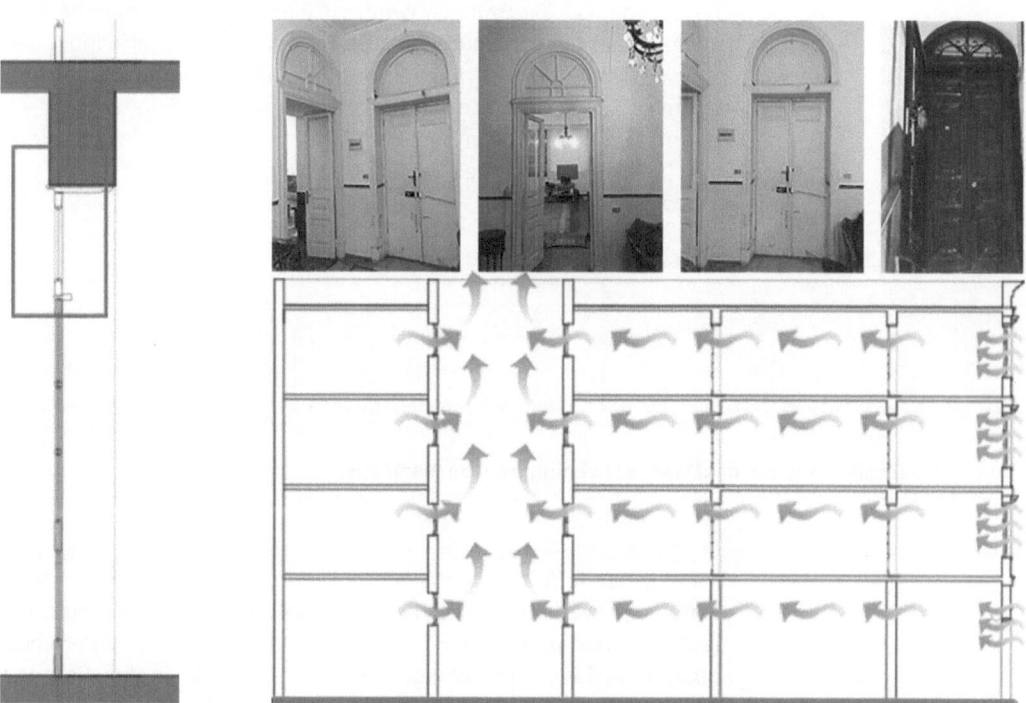

Figure 7.28 Detailed section, plan, and photographs showing the transom windows opening strategy for enhancing airflow.
(Ahmed K. Taher)

A proposed solution is to restore the existing transom windows to improve internal space connectivity and enhance ventilation – see Figure 7.28. This would be done on a per-apartment basis to maintain privacy and address safety concerns. This restoration would not alter the building's heritage characteristics but would instead return it to its original state. As a fundamental parameter for improving natural ventilation, it would serve as a baseline for comparison with other potential interventions.

The computational study cross-compared the effects of the selected measures for the natural ventilation retrofit conducted to those simulated in order to observe the changes in natural ventilation performance in the case study building, thus identifying the effectiveness of the retrofit solution. The CFD pre-simulation, solver settings, and post-processing employed for all these simulations were very similar, all based on the findings of the previous step method. These parameters were CFD simulations for the same urban domain of the surrounding context for the wind direction from the north-west. The testing scenario uses a building model matrix, the building model design, computational modelling of the testing, and boundary conditions for the CFD simulation used.

The simulation quantifies the effects of creating cross-ventilation through the connection of internal spaces, via the restoration of the existing transom windows within the detailed floor plan according to the apartment's separation, and reversing modifications applied by the occupants. The transom windows were located above the internal doors between the outer spaces (S1) and the internal living spaces (S2) see Figure 7.29, with a height 2.6 m above the floor plan and a height of 0.6 m. The 3D model was modified according to the tested parameter. The solution geometry was simplified while attaining a high level of accuracy see Table 7.5 and Figures 7.30, 7.31, 7.32, 7.33, 7.34, 7.35.

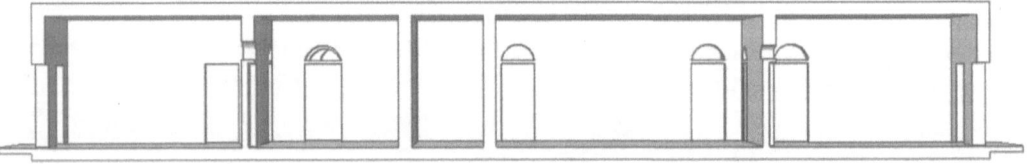

Figure 7.29 Detailed section of 3D model opening transom windows.
(Ahmed K. Taher)

Overall, the airflow patterns within the internal spaces have significantly improved due to the implementation of Scenario 1. This is particularly evident in the S2 zones, which previously experienced minimal airflow. The average airflow magnitude in the S1 zones increased from 0.26 m/s to 0.53 m/s, while in the S2 zones it rose from 0.08 m/s to 0.39 m/s.

Table 7.5 Internal airspeed inside the detailed floor plan with connected spaces

Zone		Inlet airspeed (m/s)	Internal airspeed (m/s)		Average airspeed (m/s)
			max	min	
S1	1	2.1	1.2	0.94	1.42
	2	0.74	0.44	0.21	0.48
	3	0.42	0.35	0.30	0.36
	4	0.61	0.52	0.38	0.49
	5	0.48	0.23	0.12	0.29
	6	0.36	0.42	0.14	0.30
	7	0.51	0.43	0.16	0.36
	8	0.55	0.55	0.23	0.44
	9	0.66	0.26	0.20	0.38
	10	2.08	0.66	0.26	1.00
	11	0.42	0.34	0.31	0.36
	12	0.46	0.16	0.13	0.25
	13	0.31	0.25	0.23	0.26
	14	0.53	1.12	0.42	0.69
	15	0.81	0.47	0.30	0.51
	16	0.90	1.91	0.55	1.12
	17	0.34	0.38	0.26	0.33
S2	1	-	0.70	0.53	0.62
	2	-	0.59	0.26	0.43
	3	-	0.20	0.10	0.16
	4	-	0.33	0.26	0.33
	5	-	0.07	0.05	0.07
	6	-	0.51	0.36	0.44
I	1	-	1.09	0.59	0.85
	2	-	0.81	0.23	0.53

(Ahmed K. Taher)

Based on the categorization of internal spaces and the monitoring points established in this chapter, we analysed the different spaces in the detailed floor plan to assess their airflow patterns. The S1 spaces, as identified in the floor plan, are the primary sources of airflow within the internal spaces.

For the external spaces (S1) overlooking Sizostris Street, the airflow patterns and magnitudes remained unchanged for S1(1) and S1(7) due to the lack of transom window connections. However, the other S1 spaces experienced increased average internal airflow and inlet

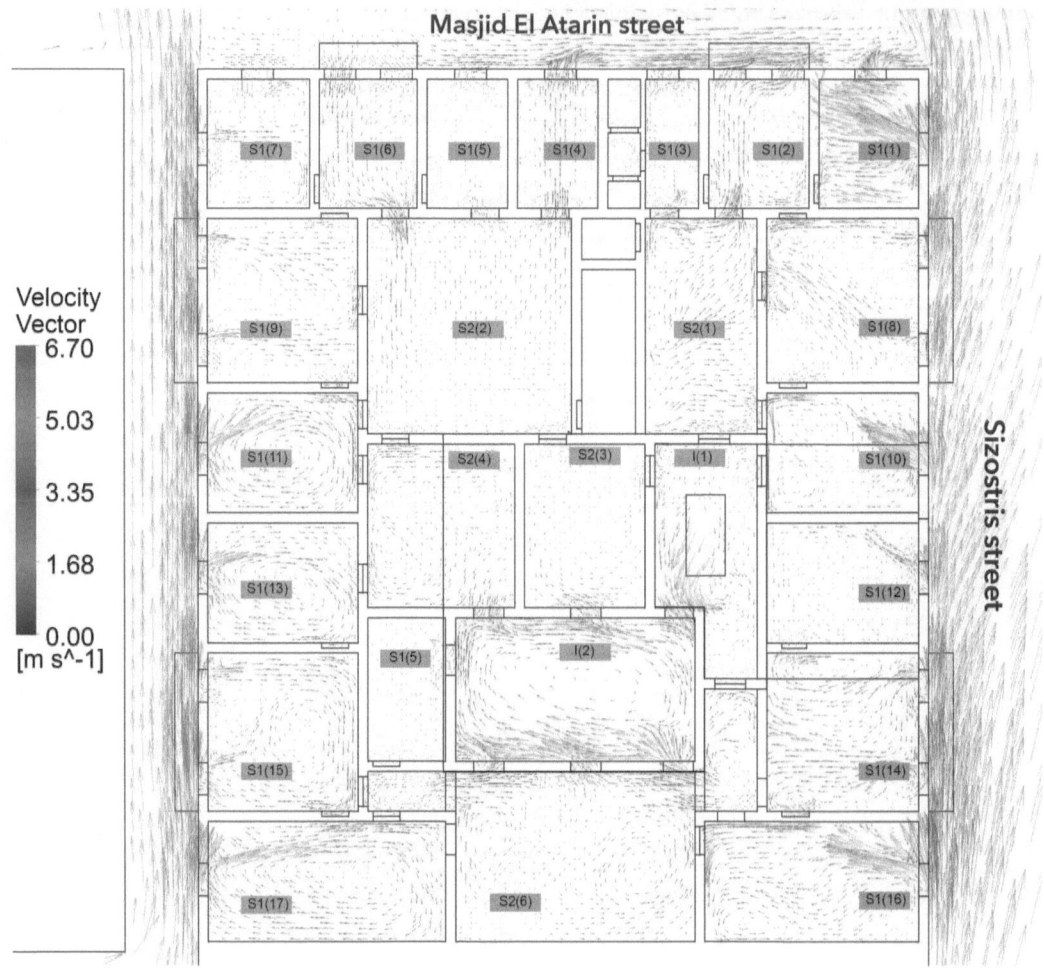

Figure 7.30 CFD simulation showing airflow pattern inside the detailed floor of the building after connecting spaces.
(Ahmed K. Taher)

airspeed. For example, S1(10) saw a significant increase in average internal airflow from 0.19 m/s to 1.0 m/s, with inlet airspeed reaching 2.1 m/s. The external spaces on the SW façade (overlooking Masjid el Atarin Street) also exhibited improved airflow. All of these spaces experienced increases in average internal airspeed and inlet magnitude. Similarly, the external spaces on the SE façade (overlooking the small alley) showed improved airflow patterns. Each space recorded an increase in average internal airspeed and inlet magnitude.

The internal spaces (S2) airflow patterns and magnitude are as follows: The S2(1) space is ventilated from the connection from the S1(8), S1(10), S1(2), and S1(3) spaces having an average internal airflow magnitude increase from 0.14 m/s to 0.62 m/s. The S2(2) space is ventilated from the S1(9), S1(11), S1(4), and S1(5) having an internal

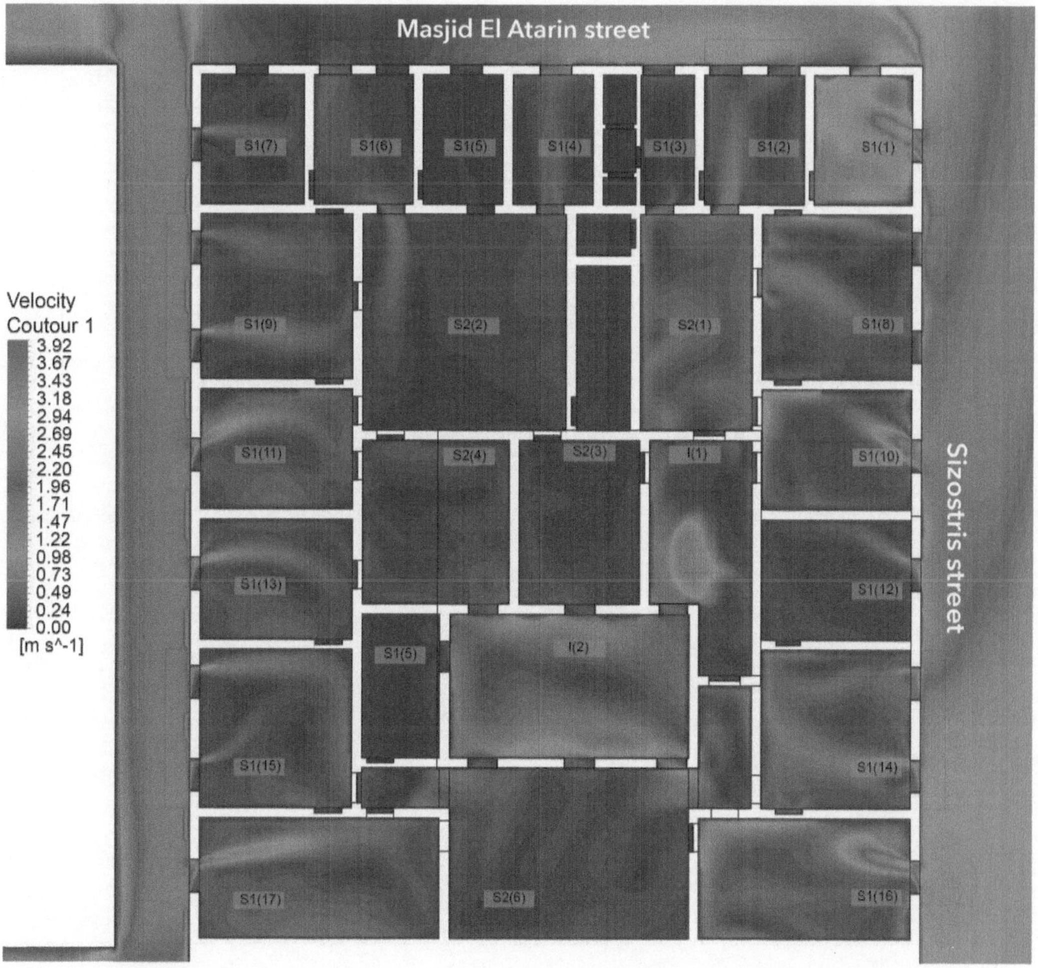

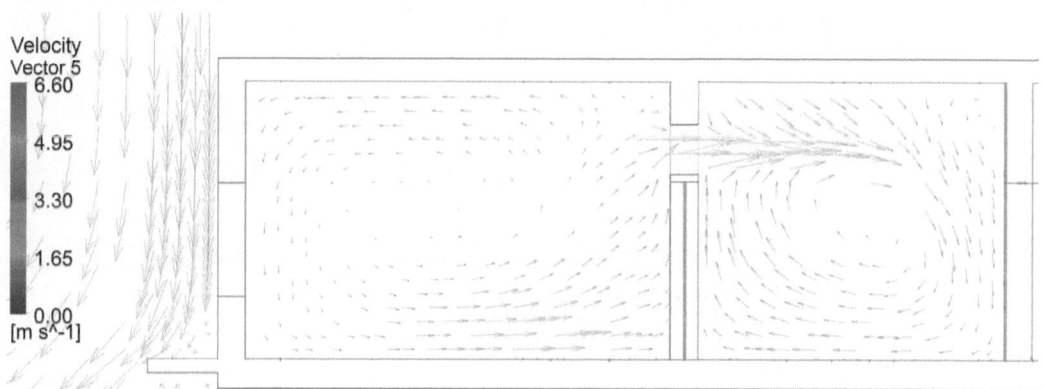

Figure 7.31 CFD simulation showing airflow speed profile of the detailed floor of the building after connecting spaces.
(Ahmed K. Taher)

Figure 7.32 Plan of the airflow patterns as a result of connecting the S1 and S2 spaces.
(Ahmed K. Taher)

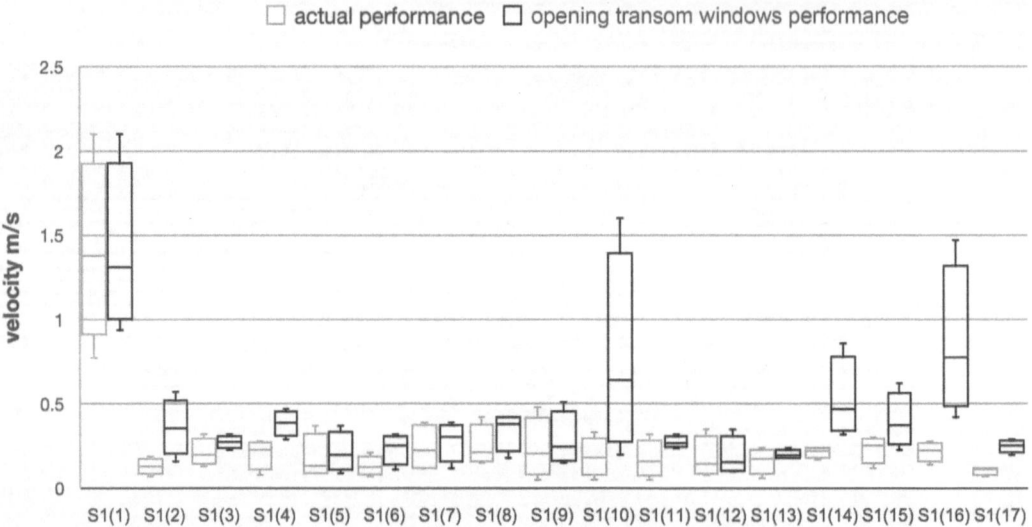

Figure 7.33 Graph showing average velocity m/s for the S1 spaces after connecting spaces.
(Ahmed K. Taher)

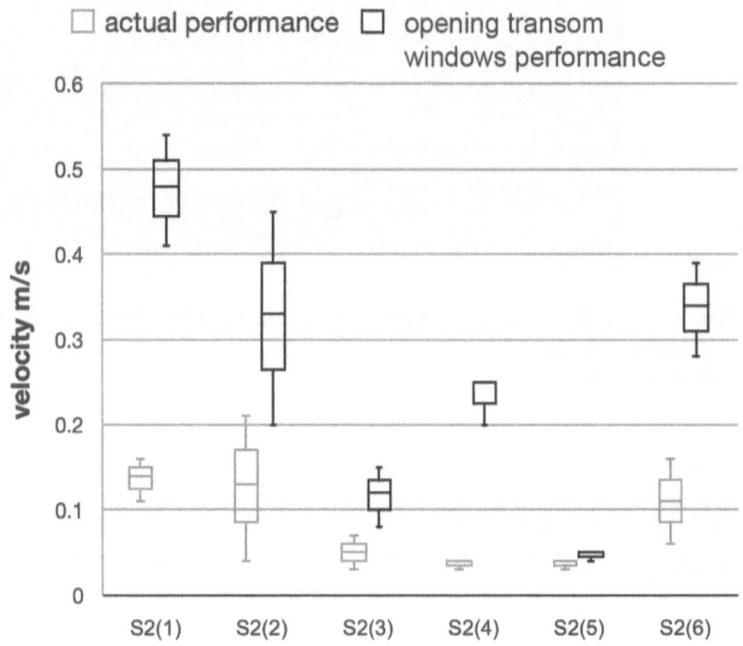

Figure 7.34 Graph showing average velocity m/s for the S2 inner spaces after connecting spaces.
(Ahmed K. Taher)

☐ actual performance

☐ opening transom
windows performance

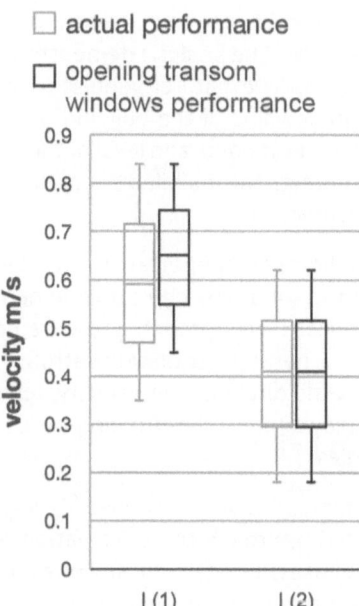

Figure 7.35 Graph showing average velocity m/s for the inner shafts' spaces after connecting spaces.
(Ahmed K. Taher)

air speed increase from 0.13 m/s to 0.43 m/s. The S2(3) is connected to S2(2) and inner courtyard with an average internal airflow magnitude increase from 0.05 m/s to 0.16 m/s. The S2(4) is connected to S1(11), S1(13), S2(2), and inner courtyard with an average internal airflow magnitude increase from 0.04 m/s to 0.32 m/s. The S2(5) is connected to the inner courtyard with no transom windows available. Airflow magnitude almost stayed the same at 0.05 m/s. The S2(6) is connected to S2(16), S2(17), and the inner courtyard with an average internal airflow magnitude increase from 0.11 m/s to 0.44 m/s.

The inner shafts, comprising the inner courtyard and staircase, acted as a negative suction force throughout the building. This helped to draw air from the outer spaces into the internal spaces. The average airflow magnitude in the inner shafts increased from 0.59 m/s to 0.84 m/s.

In summary, the implementation of opening transom windows and connecting internal spaces, has led to a substantial enhancement in airflow patterns and magnitude within the internal spaces. This is a positive development for the overall air quality and ventilation of the building.

Conclusion

This chapter presents a comprehensive framework for natural ventilation retrofit in heritage buildings. The framework emphasizes the importance of understanding the heritage context and the building's current conditions to achieve effective retrofit solutions in a hot humid maritime climate.

Through a retrofit approach, the effectiveness of passive retrofitting measures was quantified. The analysis demonstrated the potential of these measures to improve natural ventilation performance while preserving the heritage value of the building. The chapter considers factors such as the heritage value and level of the building, the impact on the heritage context, the building's structure, and the privacy requirements of occupants.

The study also highlights the importance of understanding the heritage context of Alexandria. The city's unique heritage fabric, influenced by its historical development, presents both opportunities and challenges for a natural ventilation retrofit. The city's layout, building typology, wind direction, air velocity, temperature, and relative humidity all play a role in determining the suitability of natural ventilation strategies.

The proposed framework can be applied to any type of building requiring a balanced approach to conservation and ventilation. It consists of three phases: Pre-retrofit activities, identifying retrofit options, and implementation and post-intervention evaluation. The framework aims to optimize the process of providing indoor thermal comfort while respecting the existing characteristics, subsystems, and climatic considerations of the heritage building.

A representative case study was analysed using CFD simulation to investigate airflow patterns and magnitudes within the building. The results revealed several key findings. Poor internal airflow: The average internal airflow magnitude within the internal spaces was only 0.19 m/s, indicating a significant lack of ventilation. Better external airflow: The external spaces, with direct exposure to the external environment, generally had higher airflow magnitudes (0.26 m/s). However, their performance varied depending on the number of openings. Insufficient ventilation in living spaces: The internal living spaces experienced very poor ventilation due to their complete separation from the external environment and inner courtyard. Airflow within these spaces was minimal, highlighting the challenges of achieving comfort ventilation. These findings indicate unacceptable indoor comfort conditions. The poor ventilation is primarily attributed to factors such as occupants' behaviour and modifications to the building's original design, including blocked upper openings that hinder cross-ventilation and the stack effect.

In conclusion, the framework presented in this chapter provides a valuable tool for enhancing natural ventilation in heritage buildings. By carefully considering the heritage context and implementing appropriate retrofit measures, it is possible to improve indoor air quality and comfort while preserving the building's historical integrity.

Reference List

Alex-Med. (2008). *Alexandria heritage catalogue 2007*. Alexandria.

Awad, M. (2010). *Conversation and rehabilitation in Alexandria's city center. Gamal Abdel Nasser Avenue and Salah Salem Street*. Alexandria Preservation Trust.

Awad, M. F. (1990). Italian influence on Alexandria's architecture (1834–1985). *Environmental Design: Journal of the Islamic Environmental Design Research Centre, VIII*, 72–85.

Blocken, B. (2015). Computational Fluid Dynamics for urban physics: Importance, scales, possibilities, limitations and ten tips and tricks towards accurate and reliable simulations. *Building and Environment, 91*, 219–245.

Franke, J., Hellsten, A., Schlunzen, H. A., & Carissimo, B. (2010). The Best Practise Guideline for the CFD simulation of flows in the urban environment: An outcome of COST 732. Published Online: February 8, 2011 pp. 419–427 https://doi.org/10.1504/IJEP.2011.038443 accessed 21.2.25

Haag, M. (2008). *Vintage Alexandria: Photographs of the city, 1860-1960*. American Univ in Cairo Press.

NOUH. (2010). *Principles and standards of urban harmony for heritage and special value buildings and areas. E. M. o.* Culture.

Pallini, C. (2006). Italian architects and modern Egypt. In *Studies in architecture, history and culture*, 39–50. Boston: The Aga Khan Program for Islamic Architecture at MIT.

Reid, D. M. (2003). *Whose Pharaohs?: Archaeology, museums, and Egyptian national identity from Napoleon to World War I*. Univ of California Press.

Santamouris, M., & Asimakopoulos, D. (1996). *Passive cooling of buildings*. Earthscan.

Shalaby, H. M., Sherif, A., & Altan, H. (2017). The Impact of Urban Fabric on Natural Ventilation for the City of Alexandria. In *Towards Sustainable Cities in Asia and the Middle East* Proceedings of the 1st GeoMEast International Congress and Exhibition, Egypt 2017 on Sustainable Civil Infrastructures Editors: John Calautit, Fernanda Rodrigues, Hassam Chaudhry, and Haşim Altan.

Turchiarulo, M. (2009). Building Styles brought to Egypt by the Italian Community between 1850 and 1950: The Style of Mario Rossi. *Proceedings of the Third International Congress on Construction History* [3 Volumes] Brandenburg University of Technology Cottbus, Germany, 20th–24th May 2009.

Weather and Climate. (2019). https://weather-and-climate.com/average-monthly-Rainfall-Temperature-Sunshine,Alexandria,Egypt

Yehia, M. A. (2006). *In Search of a local architectural language the case of Alexandria Egypt*. Alexandria University.

8

RETROFIT CASE STUDIES

An Historic University Accommodation Building and a Traditional Solid Walled House in Ireland

Peter Cox

Case Study 1: Historic Accommodation Buildings Project: The Rubrics Building, Trinity College Dublin 2019–2023

Figure 8.1 West elevation prior to the Conservation Works.
(Carrig Conservation)

DOI: 10.4324/9781003527404-11

Figure 8.2 Aerial view of east elevation prior to the Conservation Works.
(Carrig Conservation)

Design Team: Pascall & Watson Architects, Carrig Conservation, Passivate Building Energy Consultants and AECOM Engineers

Client: Trinity College Dublin

Building Chronology

1699–1705 The Rubrics Building (Figures 8.1 and 8.2) was built at Trinity College Dublin. Its design has been attributed to Thomas Burgh who designed the Long Library.

1838–1840 The ends of the range were demolished.

1894 The building received a new curvilinear-gabled roof and red-brick façade.

Westmoreland Green slates were imported from Cumbria. Steel roof trusses were sourced from Scotland. Portmarnock brick from north Dublin was used to clad the front west-elevation.

c.1976 The end gables were reconstructed in concrete block with red-brick cladding due to apparent bowing or instability.

Retrofit Strategy

The need to retrofit was driven by Trinity's brief to achieve modern performance standards in a historic building while respecting the heritage and continuing its use as accommodation. Working with Passivate Building Energy Consultants, a series of in-situ, lab-based and desk-based measurements and analyses were undertaken to obtain a thorough understanding of the existing building physics before work began. The condensation and thermal bridge risk analysis was done to minimise the risk of unintended consequences post-retrofit. The following building Fabric & Environment Assessments were undertaken:

- Brick porosity testing
- Petrographic analysis
- Indoor Air Quality (IAQ) monitoring
- In-situ U-value measurements
- Condensation Risk Assessment
- Thermal Bridge Analysis

Energy Conservation Works

Having reached the end of its life, the natural slate was removed from the roof. An exemplar approach was taken to insulating the roof at rafter level. Wood fibre sarking board was laid outside the rafters to minimise thermal bridging and vapour permeable insulation was friction fit between the rafters – see Figure 8.3.

Following the application of insulation, new hand cut Irish slates from Valentia were installed – see Figure 8.4. Existing rainwater goods have been restored on a like-for-like basis. Dense cementitious pebbledash, installed in the mid-20th century, has been removed from the rear east-facing elevation – See Figure 8.5. A new insulating cork lime render has been applied and finished with a roughcast lime render and a breathable water-based rain repellent.

All windows throughout the building are single-glazed timber frame units. These were removed, restored and draughtproofed. In lieu of secondary glazing, existing shutters were restored to ensure they closed tightly. Where shutters no longer or never existed, thermal curtains will be installed. External doors will be upgraded in line with fire safety compliance and will be insulated and draught-proofed. Lime plaster was used to restore internal lath and plaster ceilings. Some vulnerable ceilings were tied back to save the early plasterwork. Timber floor joists and large oak beams have been retained and strengthened, as required. Existing floorboards have been retained for reuse. An innovative approach was taken to insulating at the ground floor – 250 mm of loose fill recycled foam glass aggregate was used to warm the subfloor void while maintaining sufficient cross-flow ventilation. Essential fire stopping measures have been carried out to prevent the spread of a potential fire. A reversible system of mineral wool fire batts has been installed between floor joists.

Figure 8.3 Thermal upgrade works to the roof.
(Carrig Conservation)

**Figure 8.4 New
Valentia slates to
the roof.**
(Carrig Conservation)

New lobby partitions, which will protect the stairs from the spread of
fire from units, can also be reversed in time, if desired. New mechanical, electrical and plumbing services have been designed for the building. These carefully chosen thermal upgrades will improve the energy
efficiency of the building enough to enable the installation of a geothermal heat pump system which was the lowest carbon heat source
for the building.

Figure 8.5 Removal of the cementitious render and new insulated external render on the east elevation.
(Carrig Conservation)

Heating Solution

![Figure 8.6 photograph]

Figure 8.6 View of east elevation and New Square prior to the Conservation Works.
(Carrig Conservation)

To select the optimum heating solution for the Rubric Building, AECOM completed an extensive feasibility study, which examined a range of heating solutions. Each heating solution was scored against a range of criteria, including energy consumption, carbon dioxide

Figure 8.7 Installation of the ground collector system.
(Carrig Conservation)

emissions, conservation impact and campus impact (see Figure 8.6). The results of this feasibility study showed that, as well as having the lowest carbon dioxide (CO_2) emissions of all of the heating options, a ground source heat pump system was the optimum heating solution for this building. The closed loop collector system for the ground source heat pump consists of 21nr 170 m deep vertical boreholes in New Square, which fire a total collector length of 3,570 m over an area of approximately 2,300 m² (see Figure 8.7). Specialist predictive modelling over a 50-year period was completed by the geothermal specialist, Geoserv, to establish the collector size. This thermal modelling was based on thermal profiles of the buildings established using dynamic thermal simulations by AECOM.

Key Features of the System

- 3570 m of geothermal collector.
- 21nr. 170 m deep boreholes over an area of 2,300 m².
- 188 kW installed heating capacity.
- Delivers 425 MWh of renewable heating annually.
- 40% improvement in building fabric thermal performance.
- 75% reduction in primary energy and CO_2 emissions for the building.

Water from the collector system, with an initial temperature range of 9–12°C is used in 3nr. 63 kW ground source heat pumps to provide fully renewable space, and domestic hot water heating for the Rubrics Building. The heat pumps also use water heat recovery to generate domestic hot water for the majority of the year. Ventilation is provided by a demand controlled continuous mechanical extract system, which extracts air from all bathrooms and kitchens to maintain excellent indoor air quality levels. Indoor Air Quality (IAQ) monitoring is provided throughout the building via wall mounted CO_2 sensors. Construction commenced August 2021 with completion in March 2023.

Case Study 2: Whole Life Carbon Assessment of Two Different Retrofit Designs for a Traditional, Solid-walled, Dwelling in Ireland

The study involved the whole life carbon assessment of two different retrofit designs for a traditional, solid-walled, dwelling in Ireland.

Introduction

The first study looks at three different scenarios:

1. Where the existing building retains the internal historic fabric of architectural heritage significance;
2. Where the existing building has previously been largely stripped of its internal historic fabric; and
3. Where the existing building is replaced with a building of the same use and footprint and compliant with current building regulations.

Analysis was undertaken using OneClick LCA, for LCA modules A1–D, for a reference study period of 60 years. The building was used as the functional unit so that results are expressed in kilogrammes or tonnes of CO_2e/building. Operational (B6) energy end-uses were estimated using DEAP, and life cycle emissions were estimated using appropriate fuel emissions and primary energy factors.

Building Description

The building in this LCA study is based on a real Victorian red brick and granite building in Dublin, constructed between 1878 and 1909. It is a two-storey over raised basement, north-facing, semi-detached dwelling (see Figures 8.8–8.11).

30.85

27.90

27.30

24.89 - first fl

21.32 - upper gfl

18.36 - lower gfl

Figure 8.8 Front elevation.
(Carrig Conservation)

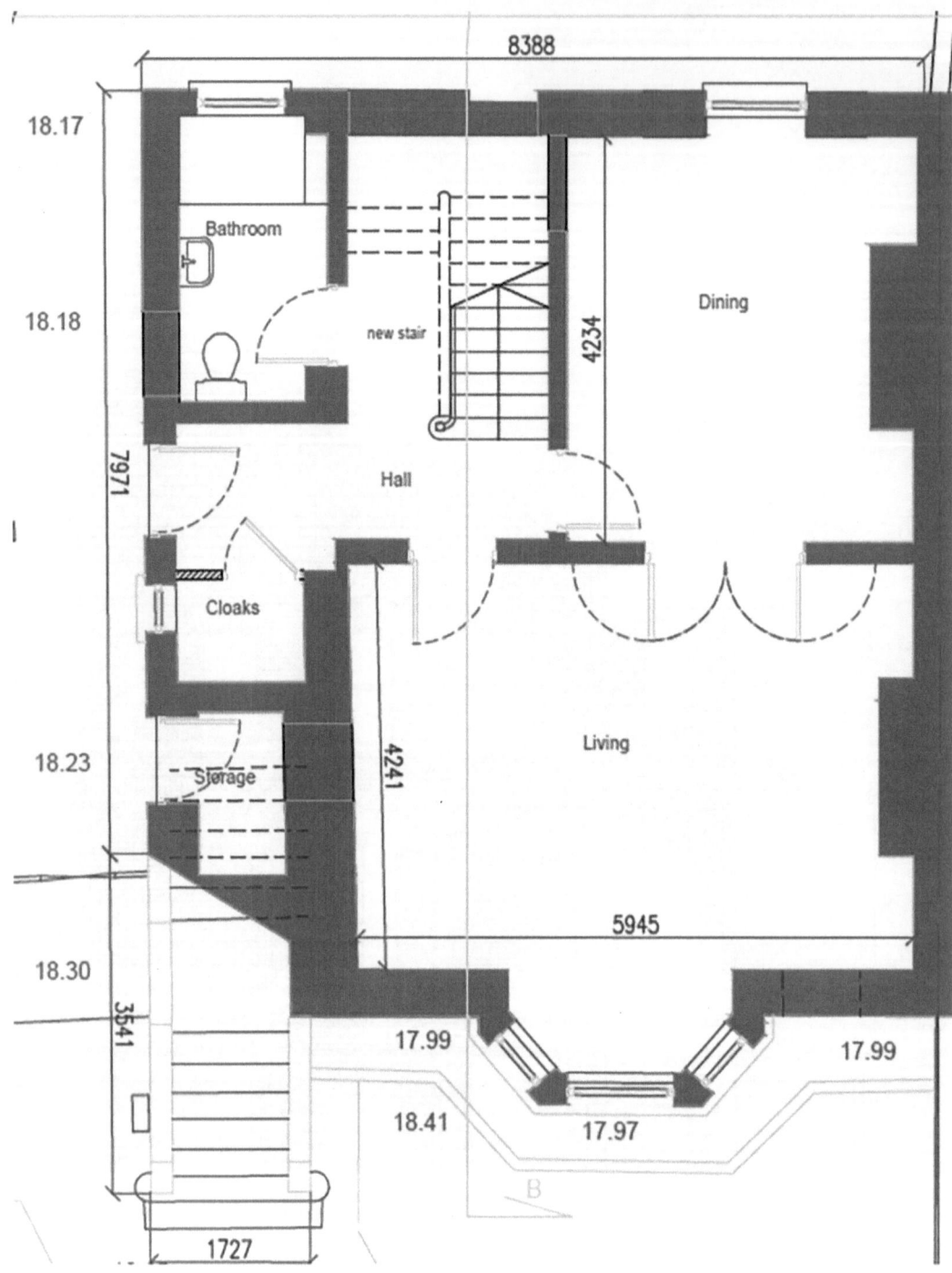

Lower Ground Floor Plan.

Figure 8.9 Lower ground floor plan.
(Carrig Conservation)

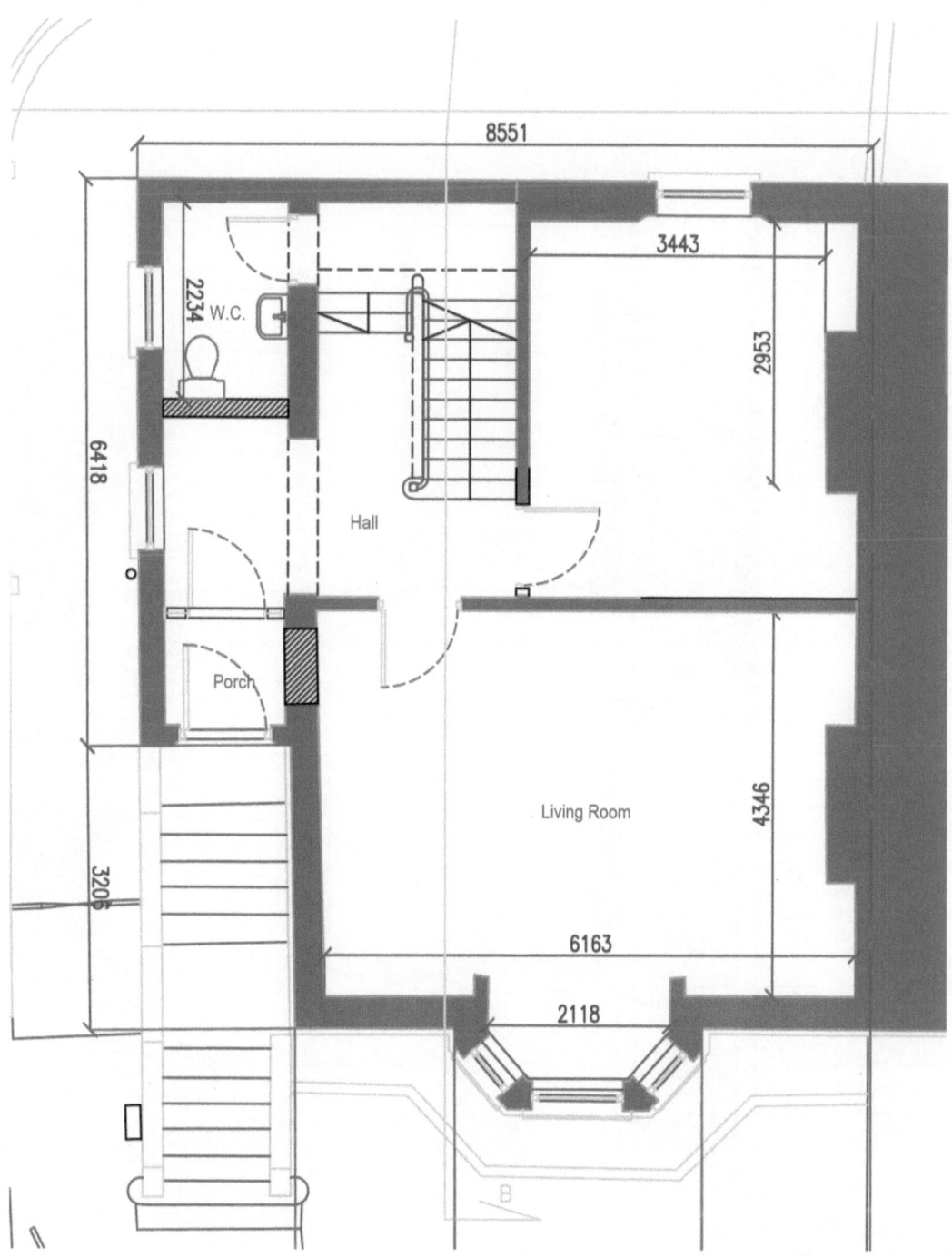

Figure 8.10 Upper ground floor plan.
(Carrig Conservation)

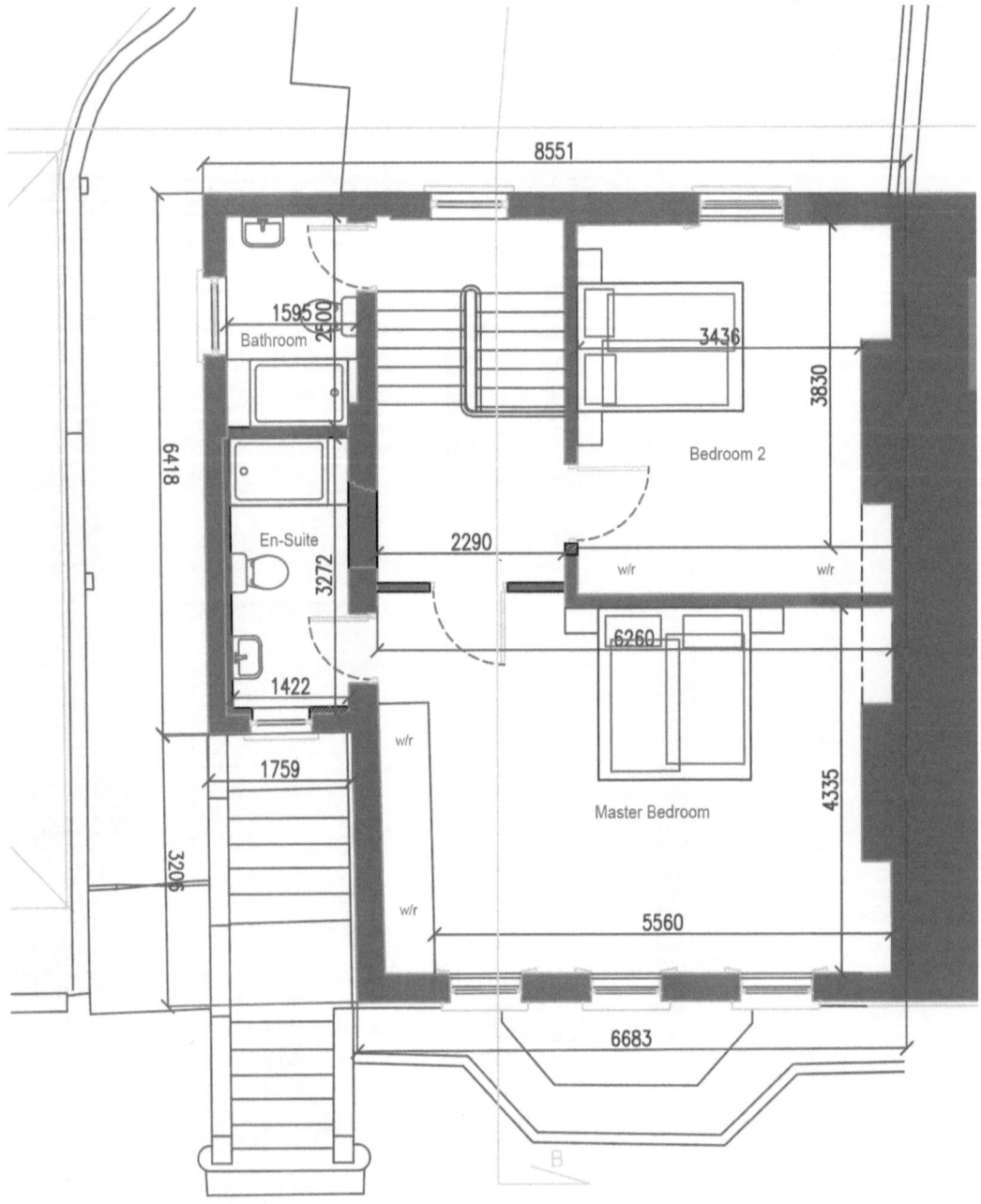

Figure 8.11 First floor plan.
(Carrig Conservation)

Table 8.1 The dimensions of the building on which the case study building is based

Levels	Length (m)	Width (m)	Height (m)	Floor Area (m²)	Volume (m³)	Perimeter (m)
Lower Ground	8.66	7.9	2.96	69	204.24	36.76
Upper Ground	8.75	8.07	3.57	67.4	240.618	35.49
First	8.88	8.09	3.01	66.5	200.165	36.35
Roof				66.5		
Totals				202.9	645.023	

(Carrig Conservation)

The structure comprises solid masonry walls, an uninsulated, ventilated suspended timber floor at ground floor, timber intermediate floors and a pitched roof with dimensions as scheduled in Table 8.1. Windows are traditional single-glazed sashes with internal shutters. Space heating would originally have been provided by means of open fireplaces and is now provided using a non-condensing natural gas-fired boiler and distributed via radiators. The property is well maintained, so that there is no water entering the structure as a result of, for example, defective roofing, guttering or render.

Design Retrofit Measures

The life cycle analysis was undertaken for three different levels of retrofit. These are described below.

Level 0 Base Case

This is the un-retrofitted 'base' case. No retrofit interventions have been undertaken and the building continues to operate as before. It represents the building without any energy-related interventions. Roofs, walls and floors are uninsulated. The existing ground floor is a ventilated and uninsulated suspended timber floor. Windows are single-glazed timber sashes, and neither windows nor doors have draught-proofing. Space heating and hot water are delivered by an existing non-condensing gas boiler with uninsulated primary pipework linked to wall-mounted radiators. Controls comprise time-only control. There is an uninsulated hot water tank and 10% of hot water needs are met using an electrical immersion element (the balance being from the gas boiler). Lighting is incandescent throughout.

Level 1 Retrofit

This represents the level of retrofit which is appropriate good practice for a building containing historic fabric of architectural heritage significance including features such as cornices and ceiling plasterwork, lime-plastered walls and timber joinery (architraves, doors, windows, skirtings, dado rails, panelling, staircases, etc.). While there is no insulation retrofitted to the walls, wood fibre insulation is provided at ceiling level.

1 The heat losses from the house mean it is not suitable for a heat pump or other electrical system.

The suspended timber floor is insulated. Draughtproofing is provided to all windows and doors, and chimneys are capped and provided with trickle ventilation. Windows are fitted with secondary glazing. There is an existing condensing gas boiler[1] with insulated primary pipework and existing wall-mounted radiators. New time and thermostatic radiator valve controls are provided. A new hot water cylinder with integrated insulation and a thermostat is provided, and 10% of hot water is met using an electrical immersion element. LEDs are fitted throughout for lighting.

Level 2 Retrofit

This represents the level of retrofit which is appropriate for a building with no surviving internal features of architectural heritage significance. Key differences with Level 1 include the insulation of walls using calcium silicate board, and a new insulated concrete ground floor with underfloor heating is introduced. Because of the better thermal performance of the building, an air-to-water heat pump is used for space heating and hot water. This is linked to new ground level underfloor heating with new wall-mounted radiators elsewhere. Controls include load/weather compensation time and temperature zone control. Table 8.2 provides a summary of key energy-related construction details associated with the different retrofit levels.

Level 3 New Build

The New Build scenario is based on the bill of quantities and design of an actual new building built in 2022 in Dublin. The quantities of materials have been amended to match the size and dimensions of the case study building. This scenario involves the demolition of the existing building, excavation works and the building of the structure from foundation. It involves fitting out and commissioning with all electrical and mechanical building systems, but excludes painting and floor finishes, as do the other scenarios.

Retrofit Design Data

The embodied carbon data used for this LCA were sourced from a paid licence for OneClick LCA and the databases that it uses. The free Carbon Designer Tool for Ireland on OneClick LCA has limited functionality as it relies on too many default values and therefore accuracy is limited. Table 8.3 and Table 8.4 show the materials that were included in each LCA module for each retrofit level and the type and source of embodied carbon data, including the database source for the generic data. Out of the 19 materials, only six had Ireland-specific data. The remainder had generic (databases) or manufacturer-specific (EPDs) embodied carbon data from Germany, France, the UK or Greece. OneClick LCA localises data for A1–A3 to an Irish market, data for A4 were altered according to average transport distances via cargo freighter from manufacturers to Holyhead, Wales. As OneClick LCA

Table 8.2 Summary of existing element details and proposed retrofit measures

Element	Level 0 Base Case	Level 1 Retrofit	Level 2 Retrofit	Level 3 New Build
Roof	Uninsulated lath and plaster ceiling	300 mm mineral wool at ceiling level	300 mm mineral wool at ceiling level	300 mm mineral wool at ceiling level
Walls	Ground floor: 340 mm stone; Upper floors: 225 mm solid brick	No additional measures	40 mm calcium silicate board onto new lime plaster internally	Cavity wall with 110 mm PIR insulation
Ground floor	Uninsulated, ventilated suspended timber floor	Existing ventilated suspended timber floor with 100 mm wood fibre insulation added	New 150 mm concrete subfloor; 100 mm PIR insulation; 75 mm screed	New 150 mm concrete subfloor; 100 mm PIR insulation; 75 mm screed
Windows and doors	Single glazed timber sashes, no draughtproofing	Draughtproofing to all windows and doors; secondary glazing	Double glazing	Low-e double glazing argon gas
Ventilation	Natural ventilation with open chimneys	Natural ventilation; capped chimneys with trickle ventilation	Mechanical ventilation heat recovery; capped chimneys with trickle ventilation	Mechanical ventilation heat recovery
Heating and hot water	Non-condensing gas boiler with uninsulated primary pipework; wall-mounted radiators; time-only control; no cylinder thermostat; uninsulated hot water tank; 10% hot water from immersion	Condensing gas boiler with insulated primary pipework; wall-mounted radiators; time and thermostatic rad valves; cylinder thermostat; integrated HW tank insulation 100 mm; 10% hot water from immersion	Air-to-water heat pump with up to 55°C flow temperature; ground level underfloor heating time and temperature zone control; cylinder thermostat; integrated HW tank insulation 100 mm	Air-to-water heat pump 45°C flow temperature; ground level underfloor heating with wall-mounted radiators elsewhere; load/weather compensation; time and temperature zone control; cylinder thermostat; integrated HW tank insulation 100 mm
Lighting	Incandescent throughout	LED throughout	LED throughout	LED throughout
Renewables	None	1.75kWp monocrystalline PV	1.75kWp monocrystalline PV	1.75kWp monocrystalline PV

(Carrig Conservation)

only allowed for the addition of one extra leg of transport the shipping of these materials through the Irish Sea was not included, however the calculation process for this would be the same. The data for all modules were obtained from databases or EPDs, therefore all end of life scenarios are provided by these sources and are not altered or reviewed for this study.

For Level 0, only modules B4–B5, B6, C and D were considered due to the level of intervention, with the existing gas boiler and radiators left in place, only requiring replacement at the end of their service lives.

Table 8.3 The materials and systems included in each LCA stage for each retrofit level

Class	Stage	Modules	Level 0 Base Case	Level 1 Retrofit	Level 2 Retrofit	Level 3 New Build
Embodied	Product	Raw material, transport, manufacture	A1–A3		Timber floor wood fibre insulation; Attic mineral wool insulation; Secondary glazing; Gas boiler; Radiators; LED lights; Mineral wool hot water tank insulation	Floor mortar; Floor steel; Floor concrete; Floor screed; Mineral wool ceiling insulation; PIR floor insulation; Calcium silicate wall insulation boards; Lime interior wall plaster; Secondary glazing; Air/water heat pump; LED lights; MVHR; Underfloor heating; Radiators	Granular fill; Drainage pipe; PIR panels; Concrete slabs with reinforcement; Radon membrane; Granular surfacing; Concrete blocks; Cement mortar; Vapour barrier membrane; Skirting boards; Interior, exterior doors; Finishing plaster; Cement plaster; Lintels; Floor screed; Rainwater goods; Gypsum plasterboard; Waterproofing membrane; Airtight membrane; Glass wool insulation; Plywood; Batten nails; Ridge tiles; Timber for roof soffit; Roof slate; Bathroom, toilet fittings – sinks, showers, faucets, baths; Sealants; Air/water heat pump; MVHR; Underfloor heating; Radiators; Electrical components – cables, sockets, switches, control box; LED lights
	Construction/Process	Transport to construction site	A4		Timber floor wood fibre insulation; Attic mineral wool insulation; Secondary glazing; Gas boiler; Radiators; LED lights; Mineral wool hot water tank insulation	Floor mortar; Floor steel; Floor concrete; Floor screed; Mineral wool ceiling insulation; PIR floor insulation; Calcium silicate wall insulation boards; Lime interior wall plaster; Secondary glazing; Air/water heat pump; LED lights; MVHR; Underfloor heating; Radiators	Granular fill; Drainage pipe; PIR panels; Concrete slabs with reinforcement; Radon membrane; Granular surfacing; Concrete blocks; Cement mortar; Vapour barrier membrane; Skirting boards; Interior, exterior doors; Finishing plaster; Cement plaster; Lintels; Floor screed; Rainwater goods; Gypsum plasterboard; Waterproofing membrane; Airtight membrane; Glass wool insulation; Plywood; Batten nails; Ridge tiles; Timber for roof soffit; Roof slate; Bathroom, toilet fittings – sinks, showers, faucets, baths; Sealants; Air/water heat pump; MVHR; Underfloor heating; Radiators; Electrical components – cables, sockets, switches, control box; LED lights.

Module	Life cycle stage	Code				
	Construction installation	A5	Timber floor wood fibre insulation; Attic mineral wool insulation; Secondary glazing; Gas boiler; Radiators; LED lights; Mineral wool hot water tank insulation	Floor mortar; Floor steel; Floor concrete; Floor screed; Mineral wool ceiling insulation; PIR floor insulation; Calcium silicate wall insulation boards; Lime interior wall plaster; Secondary glazing; Air/water heat pump; LED lights; MVHR; Underfloor heating; Radiators		Granular fill; Drainage pipe; PIR panels; Concrete slabs with reinforcement; Radon membrane; Granular surfacing; Concrete blocks; Cement mortar; Vapour barrier membrane; Skirting boards; Interior, exterior doors; Finishing plaster; Cement plaster; Lintels; Floor screed; Rainwater goods; Gypsum plasterboard; Waterproofing membrane; Airtight membrane; Glass wool insulation; Plywood; Batten nails; Ridge tiles; Timber for roof soffit; Roof slate; Bathroom, toilet fittings – sinks, showers, faucets, baths; Sealants; Air/water heat pump; MVHR; Underfloor heating; Radiators; Electrical components – cables, sockets, switches, control box; LED lights; Excavation works
Use	Use, Maintenance	B1–B2				
	Repair	B3				
	Replacement, Refurbishment	B4–B5	Gas boiler; Radiators	LED lights; Gas boiler; Radiators; Any existing materials requiring replacement during RSP such as plaster, doors, windows, rainwater goods and electrical components	Air/water heat pump; LED lights; MVHR; Underfloor heating; Radiators; Any existing materials requiring replacement during RSP such as doors, windows, rainwater goods and electrical components.	Vapour barrier membrane; Skirting board; Interior, exterior doors; Finishing plaster; Cement plaster; Waterproofing membrane; Airtightness membrane; Ridge tiles: Sealants; Air/water heat pump; MVHR; Radiators; LED lights; Electrical components – cables, sockets, switches, control box
Operational	Operational energy use	B6	Electricity, natural gas	Electricity		Electricity
	Operational water use	B7	Electricity, natural gas	Electricity		

(Continued)

Table 8.3 (Continued)

Class	Stage	Modules		Level 0 Base Case	Level 1 Retrofit	Level 2 Retrofit	Level 3 New Build
Embodied	End of life	Deconstruction-demolition	C1	Gas boiler; Radiators			
		Transport	C2	Gas boiler; Radiators	Attic wood fibre insulation; Secondary glazing; Gas boiler; Radiators; LED lights; Mineral wool hot water tank insulation	Floor mortar; Floor steel; Floor concrete; Floor screed; Mineral wool ceiling insulation; PIR floor insulation; Calcium silicate wall insulation boards; Lime interior wall plaster; Secondary glazing; Air/water heat pump; LED lights; MVHR; Underfloor heating; Radiators;	Granular fill; Drainage pipe; PIR panels; Concrete slabs with reinforcement; Radon membrane; Granular surfacing; Concrete blocks; Cement mortar; Vapour barrier membrane; Skirting boards; Interior, exterior doors; Finishing plaster; Cement plaster; Lintels; Floor screed; Rainwater goods; Gypsum plasterboard; Waterproofing membrane; Airtight membrane; Glass wool insulation; Plywood; Batten nails; Ridge tiles; Timber for roof soffit; Roof slate; Bathroom, toilet fittings – sinks, showers, faucets, baths; Sealants; Air/water heat pump; MVHR; Underfloor heating; Radiators; Electrical components – cables, sockets, switches, control box; LED lights; Deconstruction/ Demolition of existing building
		Waste processing	C3	Gas boiler; Radiators	Timber floor wood fibre insulation; Secondary glazing; Gas boiler; Radiators	Floor mortar; Floor steel; PIR floor insulation; Floor screed; Floor concrete; Secondary glazing; Air/water heat pump; MVHR; Underfloor heating; Radiators	
		Disposal	C4	Gas boiler; Radiators	Timber floor wood fibre insulation; Attic wood fibre; LED lights; Gas boiler; Radiator; Mineral wool hot water tank insulation	Air/water heat pump; LED lights; MVHR; Underfloor heating; Radiators; Mineral wool ceiling insulation; Calcium silicate wall insulation boards; Lime interior wall plaster	
	Benefits and loads beyond system boundary	Reuse, recovery and/or recycling potentials	D	Benefits of installed and replaced materials	Secondary glazing; Gas boiler; Radiators	Floor mortar; Floor steel; Floor screed; Floor concrete; Secondary glazing; Air/water heat pump; MVHR; Underfloor heating; Radiators	Drainage pipes; Concrete slabs; Concrete blocks; Cement mortar; Floor screed; Rainwater goods; Plasterboard; Batten nails; Roof slate; Bath and shower faucets; Shower heads; Sink; Pipes; Windows; Air/water heat pump; MVHR; Underfloor heating; Radiators; Electrical switch; Electrical control box; Electrical socket.

(Carrig Conservation)

Table 8.4 The source of the embodied carbon data for each material/product used in this study that did not have Ireland specific data. All other materials had Ireland-specific data

Material	Embodied Carbon Data Type	Country	Transport Route	Road Distance (km) to Dublin Port
Air/water heat pump	Manufacturer specific	France	Paris–Dublin	1057.7
Attic wood fibre insulation	Manufacturer specific	Germany	Stuttgart–Dublin	1521.2
Bathtub	Manufacturer specific	France	Paris–Dublin	1057.7
Calcium silicate wall insulation boards	Manufacturer specific	Germany	Berlin–Dublin	1695
Cement plaster	Manufacturer specific	Germany	Berlin–Dublin	1695
Copper	Manufacturer specific	France	Paris–Dublin	1057.7
Electrical control boxes, switches	Manufacturer specific	France	Paris–Dublin	1057.7
Electrical sockets	Manufacturer specific	Germany	Berlin–Dublin	1695
Floor insulation	Manufacturer specific	Germany	Munich–Dublin	1734.3
Floor screed	Manufacturer specific	France	Orleans–Dublin	1182.5
Interior, exterior doors	Manufacturer specific	France	Paris–Dublin	1057.7
LED lights	Manufacturer specific	UK	Leeds–Dublin	377.7
Lime interior wall plaster	Generic – GaBi	Germany	Berlin–Dublin	1695
Mineral wool ceiling insulation	Manufacturer specific	UK	Leeds–Dublin	377.7
MVHR	Manufacturer specific	France	Nantes–Dublin	1356.2
PV panels	Manufacturer specific	France	Paris–Dublin	1057.7
Radon membrane	Manufacturer specific	Norway	Oslo–Dublin	2333
Rainwater goods	Manufacturer specific	France	Paris–Dublin	1057.7
Roof slate	Manufacturer specific	Germany	Berlin–Dublin	1695
Secondary glazing	Generic – GaBi	Germany	Berlin–Dublin	1695
Showerhead, bath, shower faucets	Manufacturer specific	Germany	Berlin–Dublin	1695
Sink	Manufacturer specific	France	Paris–Dublin	1057.7
Skirting	Manufacturer specific	France	Paris–Dublin	1057.7
Underfloor heating	Manufacturer specific	Greece	Athens–Dublin	3775.6
Timber flooring	Manufacturer specific	Germany	Cologne–Dublin	1171.6

(Carrig Conservation)

For Level 1 and Level 2, new and removed materials were included in modules A1–A3, A4, A5, B3 and C2 but not every material had data for each life cycle module or that life cycle module was not relevant to each material. Module B7, operational water use, was not included for any Level as it was beyond the scope of this study.

The materials of the proposed retrofits which were not included in the LCA due to lack of data were: chimney caps and thermostats. It is estimated that these will only have a negligible impact on results.

Embodied – Product (A1–A3)

This stage covers impacts of a product or material that is ready to ship to construction site, including raw materials extraction, transport and manufacturing emissions. OneClick LCA also calculates biogenic carbon for relevant products but for this study, it is only reported as additional information but does not offset the stored A1 emissions or add to C3 since the product may release the carbon during disposal.

Embodied – Construction Process (A4–A5)

The transport to construction site (A4) and on-site installation/construction (A5) were standard calculations taken from the generic database data or manufacturer EPDs of each material in OneClick LCA. Where these data were not Ireland-specific, transportation data for non-Irish data were calculated by determining average road distances from the nearest city of the manufacturers to Dublin Port. OneClick LCA has a default emissions value for cargo freight which was used in this case (0.04 kg CO_2e/ton.km). The total weight of the material used was converted to tonnes by dividing by 1,000. This value was multiplied by the average travel distance and then the emissions value for the cargo freight.

Embodied – Use (B1–B5)

For this study, no data for modules B1–B2 were available, and were o for B3 therefore this stage was composed of replacement and refurbishment emissions (B4–B5) of products. For the use stage of a building life cycle, most materials are given a service life equivalent to the life of the building. The systems installed in the building have a shorter service life and were set by the generic database data or product EPDs, and displayed in Table 8.5. Existing building components such as windows and doors were not replaced at the start of the project but

Table 8.5 The service life of building systems used in this study

Element	Service Life (years)
Heat pump	22
Underfloor heating system	50
LEDs	45
Radiator system	25
Gas boiler	20

(Carrig Conservation)

they will likely need replacement within the 60-year RSP, therefore the replacement of existing doors and windows within the RSP was included.

Embodied – End of Life (C1–C4)

This stage includes impacts for processing recyclable construction waste flows for recycling (C3) until the end-of-waste stage or the impacts of pre-processing and landfilling for waste streams that cannot be recycled (C4) based on type of material. OneClick LCA calculated emissions for C2–C4 based on data from product EPDs and databases, but for the Level 3 New Build, independent calculations were undertaken to determine the demolition emissions from the theoretical demolition of the existing building. The materials that were responsible for the highest proportion of emissions/benefits for Modules C and D are presented in Table 8.6

Benefits and Loads Beyond the System Boundary (D)

This module provides transparency for the environmental benefits or loads resulting from reusable products, recyclable materials and/or useful energy carriers leaving a product system, e.g. as secondary materials or fuels or in the form of exported energy. Products such as steel, radiators, underfloor heating components and heat pumps have reuse potential at end of life and therefore may provide carbon

Table 8.6 The materials responsible for the highest proportion of emissions/benefits in Modules C and D

	Module C	Module D	Note
Level 0 Base Case	Double glazed windows, interior doors, radiators	Water filled radiators	Radiator was generic Irish-based data, scenario for Module D unspecified. Double-glazed windows and interior doors are replacement of existing windows after eventual need for replacement within the RSP. Highest emissions in C2 Waste Transport.
Level 1 Retrofit	Double glazed windows, interior doors, exterior doors, radiators	Water filled radiators	Radiator was generic Irish-based data, scenario for Module D unspecified. Double-glazed windows and interior doors are replacement of existing windows after eventual need for replacement within the RSP. Highest emissions in C2 Waste Transport.
Level 2 Retrofit	PIR insulation, underfloor heating	Underfloor heating	Highest emissions in C3 Waste Processing for PIR insulation and C2 Waste Transport for underfloor heating.
Level 3 New Build	Demolition of existing house, PIR insulation	Underfloor heating	Highest emissions in C1 Deconstruction/ Demolition for demolition of existing house and C3 Waste Processing for PIR insulation

(Carrig Conservation)

reduction benefits for another project after 60 years if they are reused. The data for this boundary were taken from product EPDs and were not calculated independently. The emissions for this boundary are presented separate to the LCA, as is required by I.S. EN 15978:2011.

Operational (B6)

The DEAP software published by the Sustainable Energy Authority of Ireland (SEAI) was used to estimate annual fuel consumption (including electricity) for each of the retrofit design options. The method also directly estimates the associated carbon dioxide emissions for the year of assessment (2023). U-values for the selected wall, window, floor and roof retrofit designs were estimated and inputted to each of the three separate retrofit models. Other key parameters relating to ventilation (e.g. chimneys, vents and draughtproofing), space and water heating systems (e.g. technology type, fuel type) were also inputted. A summary of some key DEAP input data is given in Table 8.7.

Although DEAP estimates operational emissions for the chosen assessment year, because the emissions intensity of electricity generation is forecast to decrease over the 60-year reference study period, forecast emissions for each year were calculated as the product of fuel consumption and fuel emissions factors for that year. The resulting annual energy-related emissions for each of the 60 years were summed to give the operational (B6) emissions.

Electricity emissions factors projections (see Table 8.7 and Figure 8.12) were based on Department of Public Expenditure, NDP Delivery and Reform projections (2024). Natural gas emissions factors were sourced from the SEAI (SEAI, 2023) and include the primary energy conversion factor to account for energy lost in transmission and distribution.

Estimating Life Cycle Emissions

OneClick LCA was used to carry out the calculations for life cycle emissions. For this study, the inbuilt 'Level(s)' calculation framework in OneClick LCA was used. The Level(s) life-cycle carbon version was used, which only includes global warming potential and biogenic carbon storage. Other impact categories can be selected such as ozone depletion potential, acidification and eutrophication in other tools if required. The Level(s) tool was used as it was the only tool which had an Irish adaptation available.

Analysing Results and Improving the Design

Comparative Life Cycle Emissions

The Level 0 Base Case was found to have the highest life cycle emissions at 1,010 tonnes CO_2e over the 60-year reference life, followed by Level 1, Level 3 and Level 2 at 660, 180 and 118 tonnes, respectively

Table 8.7 Selected input parameters to the DEAP operational energy and emissions model

	Retrofit Level	Units	Level 0	Level 1	Level 2	Level 3
Ventilation						
	No. of chimneys		6	0	0	0
	No. intermittent fans and passive vents		3	3	3	0
	Structure type		Masonry	Masonry	Masonry	Masonry
	Suspended timber ground floor		Natural cross ventilation	Yes – unsealed	None	None
	Windows and doors draught-stripped	(%)	0	100	100	100
Elemental U-values						
	Windows	(W/m²K)	4.03	2.19	2.19	1.33
	Doors	(W/m²K)	3	3	3	1.4
	Floor	(W/m²K)	0.65	0.24	0.15	0.15
	Walls	(W/m²K)	2.1	2.1	1.02	0.18
	Walls (type 2)	(W/m²K)	2.1	2.1	1.02	0.18
	Roof	(W/m²K)	2.3	0.13	0.13	0.13
Space heating						
	Heating system control category		2	3	3	3
	Heating system responsiveness category		1	1	1	2
Fuel data						
	Space heating – main		mains gas	mains gas	electricity	electricity
	Water heating – main		mains gas	mains gas	electricity	electricity
	Water heating – supplementary		electricity	electricity	electricity	–
	Renewables (PV)		0	1,714	1,714	1,714
Building Energy Rating (BER)			F	D1	B3	A2

1.75kWp monocrystalline PV assumed, output from PV GIS optimised azimuth/tilt

(Carrig Conservation)

(see Figure 8.12). Level 1, Level 2 and Level 3 life cycle emissions are estimated to be 65%, 12% and 18%, respectively, of the Level 0 scenario. The low Level 2 emissions are largely due to the lower fabric losses and use of an electrical heat pump, the electricity for which becomes very low carbon within the building's retrofit lifetime.

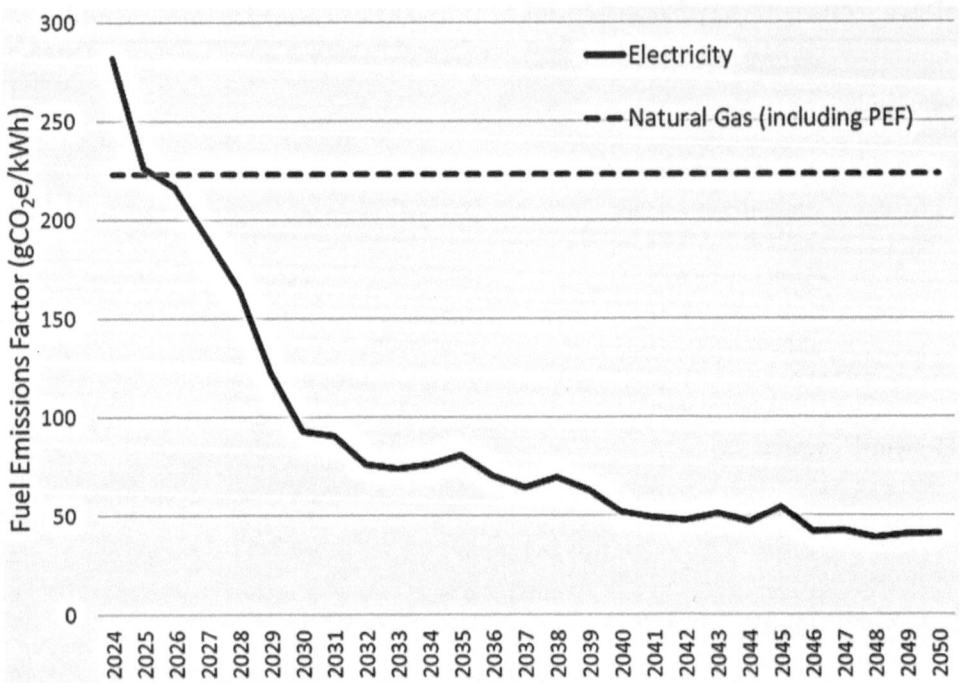

Figure 8.12 Graph of end-use electricity and natural gas CO_2e emissions factor projections up to 2050; electricity emissions after 2050 are assumed to be remain at the 2050 projected value of 40.9gCO_2e/kWh; natural gas factor accounts for primary energy factor.
Sources: SEAI (2024) and DPER (2024) (Carrig Conservation)

Level 3 has the highest embodied emissions overall (152 tCO_2e) due to the higher quantity of materials needed for a new build and the demolition emissions of the existing building. Of the retrofit options, Level 2 has the higher embodied emissions both at product and construction stages (A1–A5). This is due to the greater use of upfront insulating materials, mechanical and electrical plant, and the installation of a concrete floor. The replacement of the plant over the period also increases embodied impacts. Total embodied emissions for the Level 2 option are 54.3 tonnes, followed by Level 1 and Level 0 at 23.8 and 13.3 tonnes, respectively (see Figures 8.13 and 8.14). These lower Level 0 and Level 1 embodied emissions are due to less materials being used upfront and the use of gas boilers, rather than heat pumps, which have lower embodied emissions when being replaced.

Level 3 and Level 2 have the greatest benefits beyond the system boundary (D), at −15.3 and −14.9 tonnes CO_2e, respectively, compared to −5.8 and −6.2 for Levels 0 and 1. The improved benefit is associated with the greater use of recyclable metals in Level 2 and Level 3, in slab reinforcing (steel) and the heating systems such heat pump and underfloor heating. With regard to life cycle operational emissions, Level 0 is highest at 997 tonnes, followed by Level 1 (636 tonnes), Level 2 (64 tonnes) and Level 3 (28 tonnes) (see Figure 8.15). In all

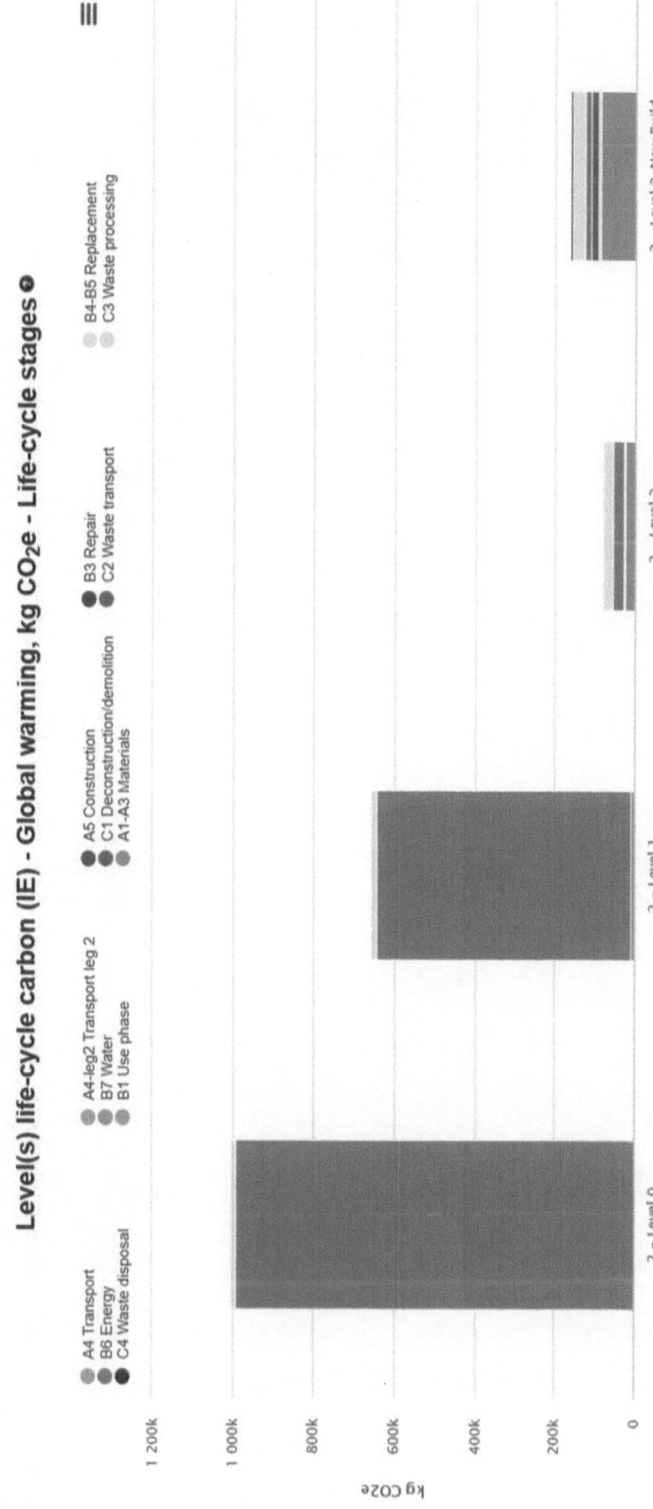

Level(s) life-cycle carbon (IE) - Global warming, kg CO$_2$e - Life-cycle stages ⊕

A4 Transport ● A4-leg2 Transport leg 2 ● A5 Construction ● B3 Repair ● B4-B5 Replacement
B6 Energy ● B7 Water ● C1 Deconstruction/demolition ● C2 Waste transport ● C3 Waste processing
C4 Waste disposal ● B1 Use phase ● A1-A3 Materials

kg CO2e

2 – Level 0   2 – Level 1   2 – Level 2   2 – Level 3–New Build

Figure 8.13 Whole Life Carbon emissions broken down by life cycle stage for each retrofit level.
(Carrig Conservation)

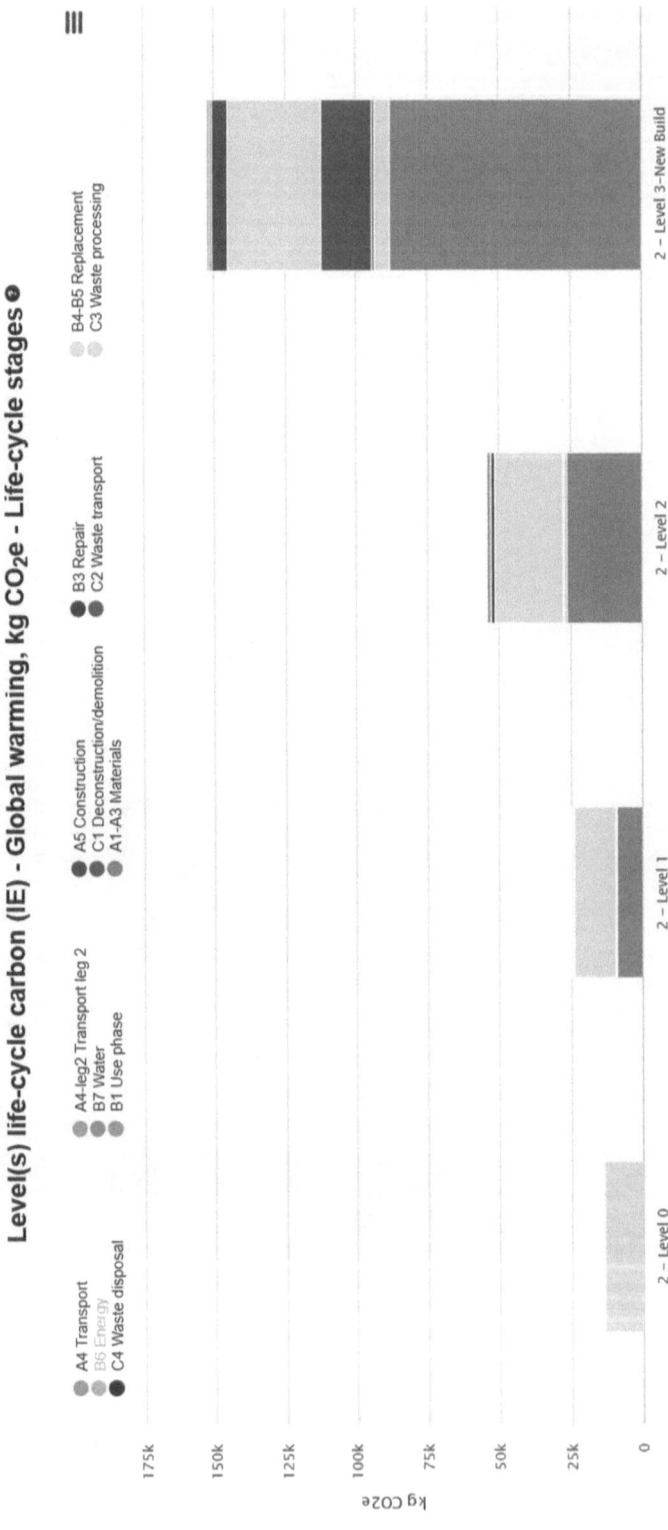

Figure 8.14 Embodied Carbon Emissions by life cycle stage and module for each of the retrofit options (excludes operational carbon emissions). (Carrig Conservation)

cases emissions are dominated by space heating, followed by hot water, with only a small contribution from lighting and small power.

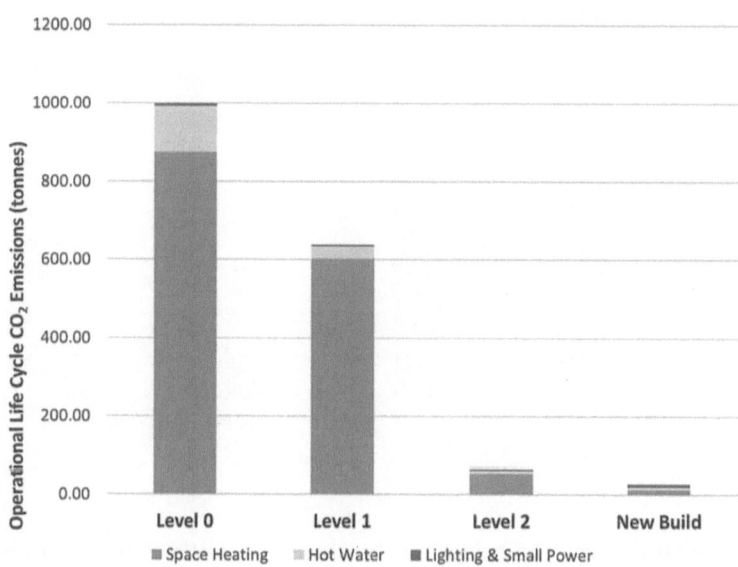

Figure 8.15
Breakdown of life cycle operational emissions by energy end use.
(Carrig Conservation)

Figure 8.15
Breakdown of life cycle operational emissions by energy end use.
(Carrig Conservation)

The primary energy use (the total quantity of raw energy inputs) per m² of floor area, as reported by DEAP, is shown in Figure 8.16. Figures are provided for each retrofit level as well as for the new-build option. These are annual 2023 emissions (i.e. not life cycle emissions), so utilise electricity emissions factors for that year. This explains why Level 2 and 3 operational emissions are relatively higher (compared to levels 0 and 1) than for the life cycle estimates above.

Hot Spot Analysis

Level 0 Base Case: operational emissions dominate and account for 99% of all emissions. This is due to no retrofit emissions, the poor thermal performance of the building fabric and need to use a fossil-fuelled heating system, which will continue to produce carbon emissions at the same rate over the life cycle.

Level 1 Retrofit: again, operational emissions dominate at 96% life cycle emissions. The reasons for this are the same as for Level 0.

Level 2 Retrofit: here, embodied emissions represent a large fraction of life cycle emissions, accounting for 46% of all emissions. This is due to the fact that the fabric has been significantly upgraded, which has two important benefits: first, it results in lower heat losses, and therefore lower energy requirements and emissions over the life cycle; and second, it reduces the heat loss indicator (or heating demand) to 2.8 W/m²K (or 59.8 W/m²), a level where it may be suitable for an electric heat pump (which has lower life cycle emissions than a gas boiler),

Figure 8.16 Primary energy use per retrofit level.
(Carrig Conservation)

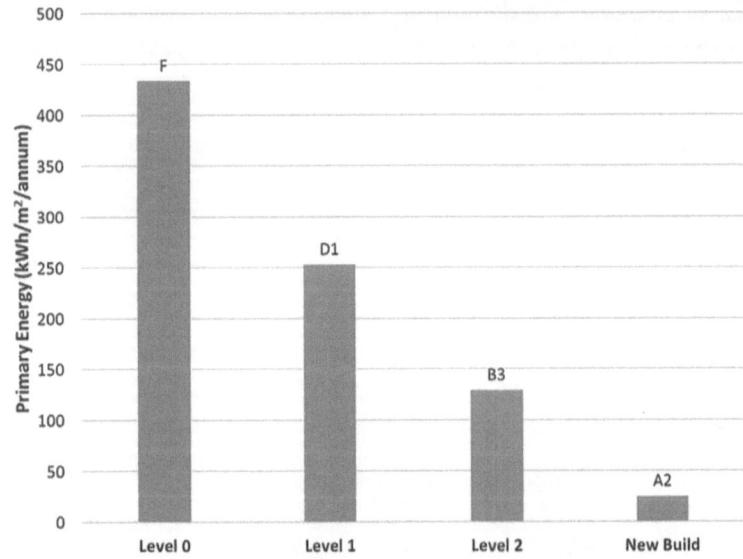

albeit marginally. A heat pump threshold of 2.3 W/m²K (50 W/m²) is currently accepted. However, there is debate whether values as high as 100 W/m² would be technically feasible, although at this level, attractiveness to homeowners may be low due to high operating costs. The heat loss indicators and unit area heating demands for each retrofit and the Level 3 New Build option are provided in Table 8.8. Focusing on the embodied emissions, it can be seen from Figure 8.17 that these are largely accounted for by services (mainly the heat pump system) and the concrete ground floor slab. A different mix design could be chosen for the slab, for example incorporating low-carbon cement replacements such as ground granulated blast furnace slag (GGBS) or pulverised fuel ash (PFA). Also, products with lower emissions could be sourced based on EPD data.

Table 8.8 Heat loss indicators and unit area heating demands for each retrofit option

Parameter	Retrofit Level			
	Level 0 Base Case	Level 1 Retrofit	Level 2 Retrofit	Level 3 New Build
Heat Loss Indicator (W/m²K)	5.7	4.3	2.8	1.7
Heating Primary Energy Demand per Unit Area (W/m²)	123.5	91.7	59.8	35.6
Annual Heating Demand (kWh/annum)	71,906	49,470	24,622	11,252
BER	F	D1	B3	A2

(Carrig Conservation)

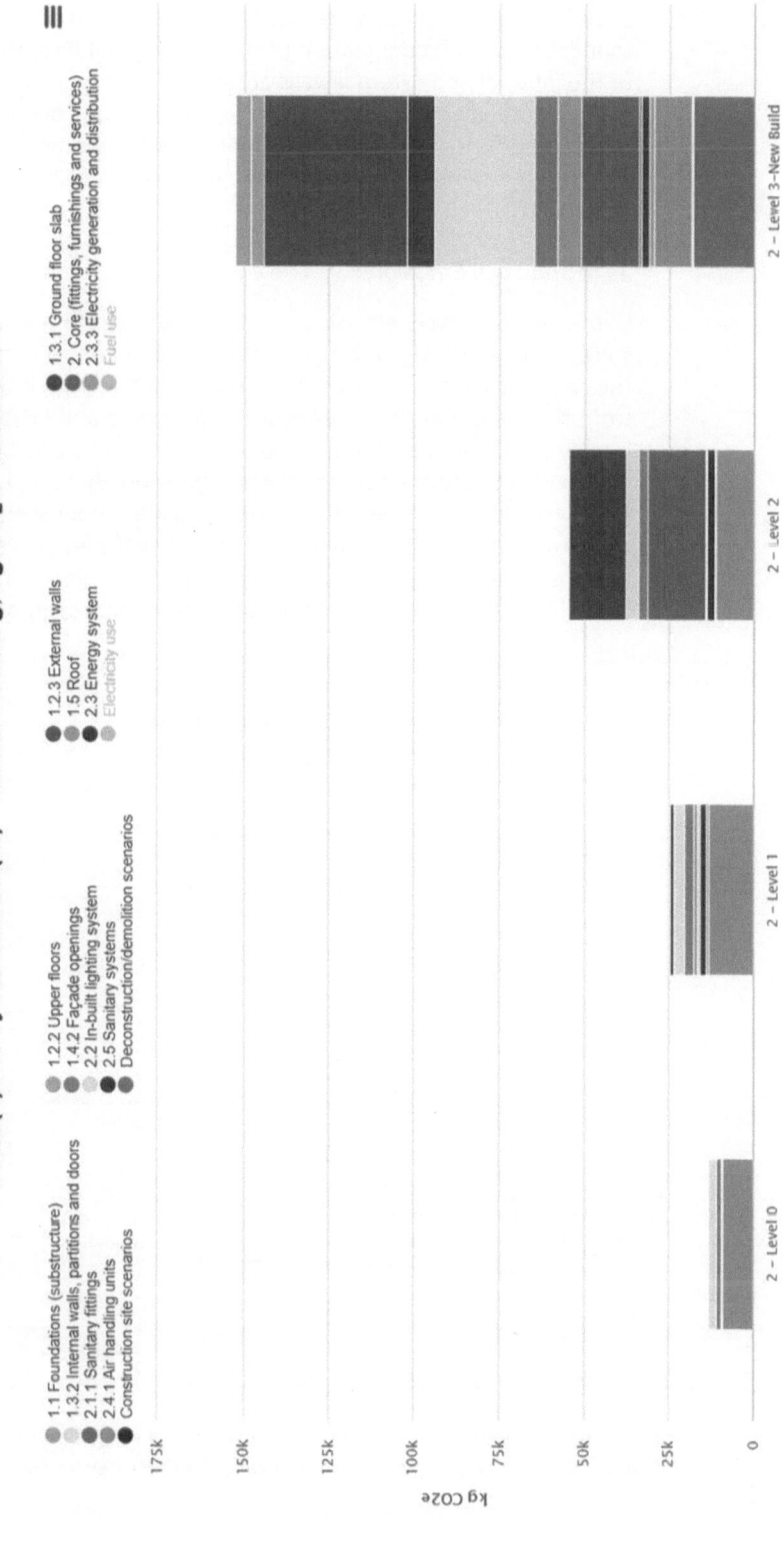

Level(s) life-cycle carbon (IE) - Global warming, kg CO₂e - Elements ⊙

● 1.1 Foundations (substructure)
● 1.3.2 Internal walls, partitions and doors
● 2.1.1 Sanitary fittings
● 2.4.1 Air handling units
● Construction site scenarios

● 1.2.2 Upper floors
● 1.4.2 Façade openings
● 2.2 In-built lighting system
● 2.5 Sanitary systems
● Deconstruction/demolition scenarios

● 1.2.3 External walls
● 1.5 Roof
● 2.3 Energy system
● Electricity use

● 1.3.1 Ground floor slab
● 2. Core (fittings, furnishings and services)
● 2.3.3 Electricity generation and distribution
● Fuel use

Figure 8.17 Embodied emissions by building element.
(Carrig Conservation)

Level 3 New Build: embodied emissions account for 85% of total emissions due to the required materials for constructing a building from foundation and the demolition of the existing building. The building of the internal and external walls accounted for a high percentage of overall embodied emissions, due to the use of concrete blocks. Alternative forms of construction, for example the use of a timber structure or cement replacements (see above), could reduce these emissions.

2030 and 2050 Emissions

Cumulative life cycle emissions for the three retrofit levels are illustrated in Figure 8.18, with the cumulative and annual emissions for the 2030 and 2050 reference years provided in Table 8.9. It can be seen from the table that annual emissions for Level 0 and Level 1 remain largely static for both years, while those for Level 2 and Level 3 fall, due to the effects of electricity decarbonisation. By 2030, Level 2 has clearly the lower cumulative emissions than all other scenarios, and this situation remains unchanged in 2050. Level 2 and Level 3 operational emissions projections are highly dependent on electricity emissions' factor projections, since both of these scenarios rely on electrical

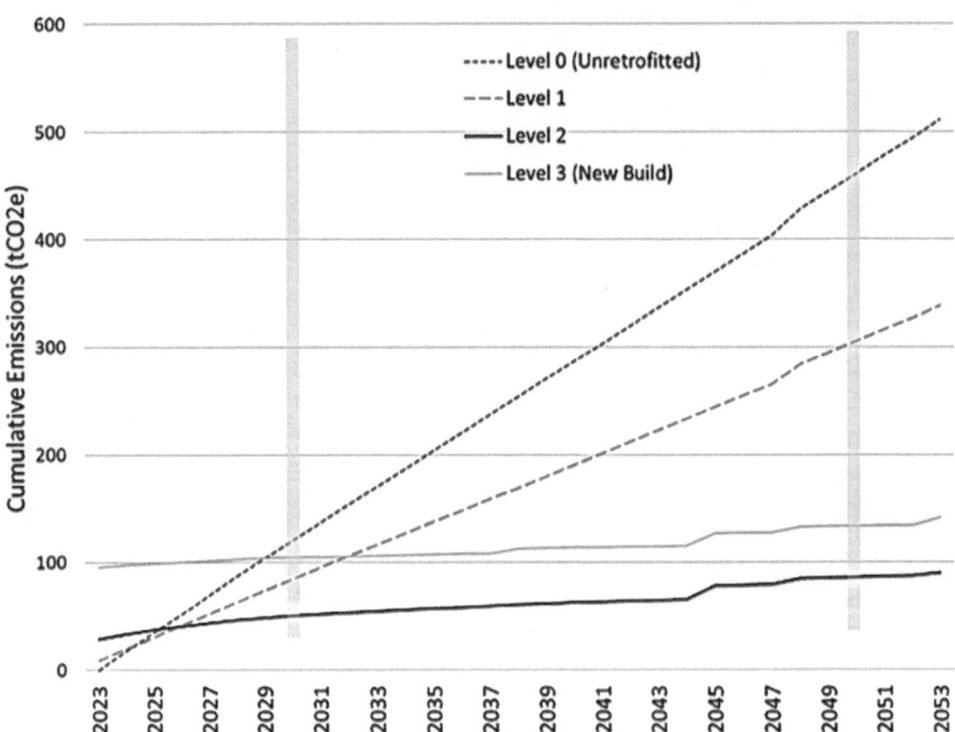

Figure 8.18 Graph of cumulative emissions over the 60-year reference study period for the retrofit options, indicating the 2030 and 2050 government target dates (grey vertical bars). (Carrig Conservation)

Table 8.9 Cumulative and annual CO_2e emissions in 2030 and 2050 for the different retrofit levels

Retrofit	Whole Life Emissions (tCO_2e)			
	2030		2050	
	Annual	Cumulative	Annual	Cumulative
Level 0 Base Case	16.7	120.0	16.5	460.9
Level 1 Retrofit	10.6	84.3	10.6	305.2
Level 2 Retrofit	1.6	50.4	0.7	86.3
Level 3 New Build	0.7	104.4	0.3	133.5

(Carrig Conservation)

forms of heating. There is some uncertainty in this regard, but this much less the case for Level 0 and Level 1, which rely to a much greater extent on natural gas, for which the emissions factor does not vary in the future. This report did not model Level 1 with an electrical heating option which might have reduced its cumulative life cycle emissions.

Conclusion

If traditional buildings are to eventually meet zero carbon targets, then energy consumption must be brought below a critical threshold whereby electrical heat pump technology can be effectively deployed at an operational cost which is acceptable to the building owner. There is, however, uncertainty about what this threshold is (both technically and economically) and how to assess it. The use of an electrical heat pump significantly lowers whole life carbon emissions, since electricity generation is forecast to have a very low carbon intensity by 2050, less than halfway into the building's reference service period of 60 years. Although it was assumed that electrical emissions would remain at 2050 levels for the remainder of the assessment period, it is possible that these will fall to zero after 2050. In this case, the Level 2 and Level 3 emissions performances would improve further.

The whole life cycle analysis report presented a review of current national and international standards, guidance, data sources and software applications relating to the retrofit of historic solid structures with the aim of minimising whole life carbon. It outlines the possible structure and content of a step-by-step guidance document for practitioners undertaking life cycle assessments of energy retrofits in traditional buildings including protected structures.

The guidance is demonstrated using a case study of a 19th-century brick dwelling, considering two retrofit options, both compared to an un-retrofitted base case and replacement with a new-build building. An important conclusion from this study is that, if it is possible to achieve practically, the deeper retrofit of traditional buildings may produce less carbon emissions than the demolition and building of a new property. Both in 2030 and 2050, the Level 2 Retrofit performs best with the lowest cumulative carbon emissions. This is also the case at the end of the 60-year reference service period, where the deeper Level 2 Retrofit whole life carbon emissions are one-third lower than those of the Level 3 New Build option.

The overall aim of the report was to propose and illustrate a step-by-step approach to assessing the embodied and operational ('whole life') carbon emissions for an energy retrofit of a traditional solid-walled building, thus providing guidance to building professionals involved in the retrofit process and to compare the embodied and operational carbon emissions of refurbishment options and replacement with a new-build of a matching footprint.

Specific objectives include:

- Based on current best practice, propose step-by-step guidance on how to include whole life carbon design into the holistic retrofit design process;
- Identify a suitable case study, apply appropriate passive (e.g. roof or wall insulation) and active (e.g. new space heating system) retrofit interventions and undertake a whole life carbon assessment using the proposed guidance;
- Detail the whole life carbon performances of the retrofit interventions;
- Compare with a replacement new-build building of similar footprint; and
- Based on the above work, undertake a gap analysis of research, guidance and data needed to facilitate holistic whole life carbon assessment in the sector in Ireland.

For the purposes of this document, traditional buildings are defined as follows:

> *Traditional buildings in Ireland generally include those built with solid masonry walls of brick, stone or clay, using lime-based mortars, often with a lime or earthen-based render finish, single-glazed timber or metal-framed windows and a timber-framed roof usually clad with slate but often with tiles, copper, lead or, less commonly, corrugated iron or thatch. In general, these were the dominant forms of building construction from medieval times until the second quarter of the twentieth century.*[2]

Finally, gaps were identified in the guidance and data resources for practitioners involved in minimising whole life carbon of a historic building retrofit. These include:

- An absence of comprehensive LCA guidance which is specific to the retrofit of traditional buildings as practiced in Ireland, and which is compliant with I.S.EN 15978:2011;

- Representative, accurate data on the energy and emissions performance of existing traditional buildings;
- A holistic LCA tool which calculates embodied carbon emissions as well as annual operational energy use and that is specific to traditional buildings;
- Limited Agrément and EPD certification for materials suitable for retrofitting traditional solid-walled buildings; and
- A lack of clarity on the suitability thresholds for deploying air source heat pumps into solid-walled traditional buildings.

It should be noted that this study focused solely on life cycle carbon, without taking into account any cost-benefit analysis. Some measures, while they may save significant carbon emissions, may not be economically feasible or widely applicable. There are more factors than purely carbon to consider when determining which retrofit options to adopt, such as cost, impact on heritage values and the impact of environmental indicators other than carbon. Therefore, this study should be considered a first step in a more holistic study to determine how to retrofit traditional buildings which balances carbon reduction targets with other key criteria.

Bibliography

DPER; Government of Ireland (2024). *Infrastructure guidelines: Supplementary guidance changes in greenhouse gas emissions in economic appraisal.* Dublin: Department of Public Expenditure, NDP Delivery and Reform. https://www.gov.ie/pdf/??file=https://assets.gov.ie/291235/6ecda5db-529b-46a3-ae82-c016857ad78a.pdf/?#page=null

Government of Ireland (2024). *Energy upgrading of traditional buildings for low embodied and life cycle emissions: Guidance and case study.* Dublin: Department of Housing, Local Government and Heritage.

SEAI Sustainable Energy Authority of Ireland (2023). *Conversion factors.* https://www.seai.ie/data-and-insights/seai-statistics/conversion-factors/

9

EMBODIED CARBON AS A PIVOT

The Case of M&S

Henrietta Billings

Introduction

Figure 9.1
Orchard House.
(Matthew Andrews for
SAVE Britain's Heritage)

DOI: 10.4324/9781003527404-12

The grounds for refusing the scheme have only grown in the 3 years since Marks & Spencer (M&S) made its planning submission, and there is now an unanswerable case for the new government to act in accelerating the industry's shift towards reusing, repurposing and extending buildings instead of demolishing and wasting them. Working with the *Architects' Journal* and sustainability experts we used this campaign to drill down into the environmental costs of demolition and to press home the well coined phrase by Carl Elefante, the former president of the American Institute of Architects, *"the greenest building is the building that already exists"* (Elefante 2007), a message that fits hand in glove with our work to save historic buildings and prevent wasteful loss.

In October 2022, SAVE squared up to M&S at a public inquiry into the retailer's proposal to demolish their flagship building on Oxford Street (see Figure 9.1) and replace it with a 10-storey office block. We led the opposition during the 2-week hearing with our brilliant barrister and expert witnesses on carbon and heritage. It was the first time sustainability and heritage were at the heart of a planning inquiry and there was significant media interest. The inquiry was seen as a major test of our disposable, knock-it-down and rebuild attitude to the built environment.

What followed next was a number of dramatic decision-making twists and turns – including a tremendous success for SAVE when the Secretary of State overturned the Planning Inspector and halted M&S' plans (Waite et al. 2023). This was followed by a procedural appeal in the High Court and the final decision then returned to the Secretary of State. At the time of writing, that decision from the new government is pending and M&S still don't have permission for their plans.

What is clear is that a consensus is emerging – that the dial is shifting away from needless demolition. Since the plans were first submitted in 2021 we've seen a marked change in interest from developers exploring retrofit schemes, a rise in the number of associated planning applications and a raft of emerging local plan policy introducing retrofit policies for new development. Developers like Seaforth Land, General Projects and Fore Partnership are raising their heads above the parapet to challenge accepted norms about demolition and rebuild, showcasing schemes such as Bleeding Heart Yard or Space House in London as imaginative, bold comprehensive retrofits of existing buildings. We are seeing architects being asked to look at buildings with fresh eyes and recent comprehensive refurbishments from London and Gloucester to York and Edinburgh show just a fraction of what is possible.

Climate Emergency

In 2019, following a recommendation from the Climate Change Committee, the government committed to a 100% reduction in greenhouse gas emissions by 2050 as compared to a baseline of 1990. This was done via the Climate Change Act 2008 (2050 Target Amendment) Order 2019. This is referred to as the net zero target and is legally binding.

The UK is a signatory to the Paris Agreement, an international treaty adopted at the United Nations climate change conference (COP21) in Paris in 2015. The target of the Paris Agreement is to limit any increase in global surface temperature to 1.5°C. It also requires countries to submit nationally determined contributions – national plans to cut emissions. The UK's target for 2030 commits the government to reducing greenhouse gas emissions by at least 68% compared to 1990 levels (HOC 2024).

In 2022 the House of Commons Environmental Audit Committee issued a report on the carbon cost of construction which highlighted the M&S scheme as a case that, *"brings the debate regarding the environmental credentials of new-build versus retrofit into public focus"* (Environmental Audit Committee 2022-23 Box 3, p. 58).

The MPs said retrofit should be prioritised and warned: *"If the UK continues to drag its feet on embodied carbon, it will not meet net-zero or its carbon targets"* (Environmental Audit Committee 2022-23 para. 72).

The report, entitled: "Building to net zero: costing carbon in construction" states that a quarter of the UK's total greenhouse gas emissions are attributable to the built environment. Embodied emissions from the demolition and construction of buildings amounts to some 40 to 50 million tonnes of CO_2 per year, more than emissions from aviation and shipping combined (Environmental Audit Committee 2022-23 para. 2). The report goes on to state:

> *The construction, demolition and excavation sector is responsible for 62 per cent of the total waste generated in the UK. It is estimated that 80 per cent of buildings currently standing will still be in use in 2050. If the UK is to meet its net zero goals the majority of these will require retrofitting to become energy efficient.*
> (Environmental Audit Committee 2022-23 para. 185)

It adds: *"There is a clear policy imperative to reduce the consumption of resources in the building and construction sector, to reduce waste material arising from demolition and replacement of existing properties, and to prioritise work to reduce emissions attributable to the built environment."* (Environmental Audit Committee 2022-23 para 186).

The importance of embodied carbon impacts when weighing up planning decisions is widely expected to increase. The government has already pledged to review the National Planning Policy Framework (NPPF) to ensure it contributes to climate change mitigation which includes a review of incentives for retrofit and the role of circular economy statements. One of the 27 priority recommendations of the

Climate Change Committee's report in June 2023 was to *"review and update the NPPF to ensure net zero outcomes are consistently prioritised through the planning system, making clear that these would work in conjunction with, rather than being over-ridden by other outcomes such as development viability"* (Climate Change Committee 2023).

Embodied Carbon

"Embodied" carbon emissions are those linked with raw material extraction, manufacture, transportation to site, and construction, maintenance and replacement of components, and those associated with the demolition of existing buildings on the site and eventual disposal. This is different from "operational" carbon emissions, linked with the energy required to run a building, such as heating and cooling, water heating, ventilation and lighting.[1]

The M&S Buildings and History

1 Definition of embodied carbon, London Plan, 2021, para 9.2.11 (Mayor of London 2021).

Figure 9.2 Detail of Orchard House showing detail of metal windows and Portland Stone façade and pilasters. (Matthew Andrews for SAVE Britain's Heritage)

Orchard House is a handsome corner building, built in 1929 by J Lyons & Co – as offices and shops (Figure 9.2). Marks & Spencer were seeking to establish a West End store under the leadership of Simon Marks, the founder's son and chairman for more than 50 years. According to the Survey of London: *"Marks was determined to open a store in Oxford Street, assuring his financial advisors that 'even if it never makes a profit, it will be a good advertisement for business'"* (Saint 2020 chapter 11, p. 8). He was right – and just two years later the Queen visited the store – an event that hit the headlines.

Orchard House is one of three buildings that make up the site of the centre of the contested proposals. The two other buildings include Neale House, a red brick 1986 office building on Oxford Street and 23–24 Orchard Street, built as an extension to the M&S building in 1968–70. Initially Marks & Spencer leased the ground floor and basement of Orchard House only, keen to establish themselves in London's West End. Their success saw them open the Pantheon store in eastern Oxford Street in 1938, and then expand into the whole of 23–24 Oxford Street in 1968–70 and to Neale House in 1986. The Marble Arch branch overtook the Pantheon and cemented its position as M&S' flagship store. All five floors remain in operational use by M&S.

Orchard House was designed by architects Trehearne and Norman who also designed a number of grand civic buildings on the Aldwych in central London. It's location next to Selfridges is a unique and special point of interest. The visual interplay between the magnificence of one of London's greatest Beaux Arts buildings and the comparative deference and dignified austerity of Orchard House is clear. The use of Portland Stone, metal windows and scale and massing are some of the obvious architectural references borrowed from its illustrious neighbour. They create a historic ensemble of exterior features which reflect the building's role as a major landmark on Oxford Street.

Orchard House is not listed and does not benefit from conservation area status. Nevertheless, in his evidence to the public inquiry on behalf of SAVE, Alec Forshaw pointed out the strong significance of the 20th-century department stores that feature along Britain's most famous shopping street. He said:

> *Oxford Street is distinguished by its succession of fine 20th century department stores which act as regular landmarks along both sides of the street. They combine, in the words of Pevsner, to produce the effect of a flotilla, sailing majestically along the street. Orchard House is one of those great galleons.*
>
> (Forshaw. A 2022 para 4.21, p. 7)

The Proposal

Under the plans submitted by M&S, all the existing buildings at 456–472 Oxford Street would be demolished to be replaced by a 10-storey building. The development would also include a two-storey basement across the whole site. The contentious design of the plans has drawn wide criticism for a number of reasons including its monolithic form, heavy and dominant cornice and impact on views from surrounding conservation areas.

The proposed building, designed by architects Pilbrow and Partners, would have retail in the basement and the ground floor, reportedly for M&S to retain as a food store, with speculative office space on the floors above. The new build submission reduces the retail by about two thirds, from around 30,000 m² existing to around 10,000 m². Even this amount of retail is not guaranteed by M&S as the planning drawings only commit to retail on part of the ground floor.

The M&S plans would also come at a high carbon cost releasing almost 40,000 tonnes of embodied CO_2 immediately. That's the equivalent of driving a typical car 99,000,000 miles (further than the distance to the sun). These are figures produced by Arup engineers, part of proposal team for the site. M&S, who lease the site from the Portman Estate, have consistently claimed the buildings cannot be retained and refurbished to provide the retail and office uses they desire.

The Campaign

SAVE first objected to the scheme in November 2021 ahead of Westminster City Council's planning committee decision where Councillors nodded through the plans. Just days before Historic England, the government's heritage advisors, announced that a listing bid for Orchard House by the Twentieth Century Society had been unsuccessful. We publicised our concerns, highlighting the huge environmental and heritage costs of the proposals – and were struck by the very high levels of public interest in the case from the start (SAVE Britain's Heritage 2021).

The Sturgis Report

As part of our campaign, in January 2022 we commissioned carbon and sustainability expert Simon Sturgis to examine the carbon impact of the proposals. His report, "Why a Comprehensive Retrofit is more Carbon Efficient than the Proposed New Build" (SAVE Britain's Heritage 2022c) found that the proposals did not comply with the UK Government's legally binding net zero legislation to reduce carbon emissions or the Greater London Authority's stated policy to prioritise retrofit. The plans also ran counter to Westminster City Council's declaration of a climate emergency. The Sturgis report also showed the buildings could be successfully brought up to contemporary environmental and retail standards. A retrofit would also produce carbon emissions, but these would be significantly less than those for the proposed new build.

Sturgis also stated that a comprehensive retrofit was an opportunity to explore a new form of architectural solution for sites like these.

Whatever its sustainability credentials, the existing buildings are valuable carbon assets already on the site eminently capable of reuse. He argued that in order to meet national, London-wide and local carbon targets we must look at the total carbon equation including a comprehensive retrofit. Sturgis highlighted that such a scheme would provide better internal layouts, modern environmental standards and additional space and would be a significantly lower carbon option compared with demolition and rebuild. The report called on M&S to undertake a detailed and comprehensive exercise to design such a scheme. He urged them to apply that same level of ingenuity and design skill that has been applied to the new-build proposal to a comprehensive retrofit scheme (Sturgis 2022).

Following a request by SAVE in April 2022, the Secretary of State Michael Gove issued an Article 31 holding direction – suspending Westminster's planning consent until the government had scrutinised the plans. At the same time we also published our report "Departing Stores: Emporia at Risk" examining the threats, challenges and future opportunities for reusing department stores including Orchard House.[2]

Open Letter

With the campaign rapidly gaining public attention, in May 2022 we published an open letter signed by leading architects, engineers, urbanists and historians urging the government to "call-in" the scheme for examination at a public inquiry (SAVE Britain's Heritage 2022a).

We received support from an impressive range of architects, engineers, developers and public figures including Julia Barfield, designer of the London Eye, and Stirling Prize winner Steve Tompkins who founded Architects Declare; academics Dr Lesley Lokko and Dr Alice Moncaster, TV personalities Kevin McCloud and Griff Rhys Jones, politicians including Duncan Baker, MP for North Norfolk who brought the Carbon Emissions (Buildings) Bill to Parliament; plus Will Arnold head of climate action at the Institute of Structural Engineers and architectural historians Andrew Saint, Alan Powers and Barnabas Calder (SAVE Britain's Heritage 2022b).

In June 2022 a 2-week public inquiry was announced – overseen by a government planning inspector appointed by the Secretary of State. The date was set for 4 months later, and we launched our most high-profile crowdfunding campaign to date, raising over £20,000 towards our legal costs.

The inquiry opened at Westminster City Hall on 25th October 2022. SAVE was represented by barrister Matthew Fraser of Landmark Chambers and three expert witnesses two tackling sustainability and one addressing heritage.

The campaign made headlines in papers from *The Telegraph* to *Time Magazine*. We were thrilled by the powerful spotlight focused on the case. The *Daily Mail* sent a reporter to the opening day, while *The Times* and *The Guardian* featured statements to the inquiry by Kristin Scott Thomas and Griff Rhys Jones. *The Guardian* ran a column by Simon Jenkins entitled: *"M&S is a shining example of how not to treat the high street – or the planet"* (Jenkins 2022). Columnist Catherine Bennett wrote in the *Observer*: *"Razing your architectural gem is a funny way to show a love for heritage, M&S"* (Bennet 2022). Michelle Ludik, from global architects HOK, wrote in the *Evening Standard* that it was *"easy to see why the demolition of a prominent and longstanding building in central London has become a matter of national interest"* (Ludik 2022).

Simon Sturgis was our expert witness on embodied carbon and sustainability. He founded Targeting Zero and is a qualified architect, and wrote our earlier carbon report on the M&S proposals. He is a member of the Construction Industry Council (CIC) Climate Change Expert Panel, and a member of the British Council for Offices Sustainability Group. He is an advisor to the EU Commission, and the Green Construction Board on sustainability. He was Special Advisor to the Environmental Audit Select Committee, was lead author for the RICS 2023 publication "Whole Life Carbon assessment for the built environment" (Royal Institution of Chartered Surveyors (RICS) 2023). This is the standard UK carbon assessment methodology.

Dr Julie Godefroy was our expert witness on whole life carbon and sustainability. She is a chartered engineer and sustainability consultant with a PhD in low-carbon buildings. She is head of sustainability at The Chartered Institute of Building Services Engineers (CIBSE), the leading authority and standard setter on building services engineering, and sustainability adviser for the National Trust's Historic Environment Group.

Alec Forshaw was our expert witness on heritage. He is a writer, planner and urban designer who was head of conservation at Islington council for over 30 years. He has successfully represented SAVE at numerous inquiries including Smithfield Market in London and Anglia Square in Norwich.

Many people made statements to the inquiry in support of SAVE's case both in person and in writing. Griff Rhys Jones[3] said: "Recycling good historic buildings should be at the heart of policy" (Hurst 2022). Architect Julia Barfield reminded the inquiry of the stark warning from the Intergovernmental Panel on Climate Change in 2018 that we have 12 years to avoid a catastrophe – now well under a decade. She said:

> What I think is at issue at this public inquiry in 2022 is: are we are acting as if there is an emergency? In my view, throwing a huge carbon bomb unnecessarily into the atmosphere – as this project proposed to do – is definitely not acting like there is an emergency.
>
> (Hurst 2022)

[3] Griff Rhys Jones, President of the Victorian Society and TV presenter (for SAVE).

Local Westminster Cllr Jessica Toale (now MP for Bournemouth West) said:

> M&S could set an example for other owners, occupiers and investors across the country by demonstrating leadership in the re-use and retrofit of heritage assets – rather than holding the community to ransom with threats to abandon the site - and in the process make a significant contribution to the country's climate goals.
>
> (Hurst 2022)

SAVE's Case: Sustainability

In his evidence, Sturgis said:

> The proposed demolition and new construction at 456 Oxford Street is in direct opposition to the government's net zero obligations and objectives and the aligned policies and commitments by all parties at all levels of decision making on this submission [Westminster City Council, the Greater London Authority]. These policies and commitments are consistently in favour of low-carbon design, resource efficiency, prioritization of retrofit and circular economic outcomes, but these policies and commitments have not been pursued for 456 Oxford Street.
>
> (Sturgis 2022 para 2.11)

He added:

> It is entirely possible, based on my appraisal of the available information, for the existing buildings to be retrofitted, reorganized and extended for a significantly lower carbon cost than the carbon cost of a new build. I consider this approach has been superficially examined by the applicants, who have in their submitted application only presented a carbon assessment of a 'light touch refurbishment'. This is an option that was always bound to fail in a comparison with the new build option.
>
> (Sturgis 2022 para 2.5, p.5)

M&S were unable to demonstrate that they had considered a deep retrofit or that retention was fully explored before the decision to demolish was cemented into the brief given to the architects. SAVE's barrister Matthew Fraser observed: "*For all that is known from this evidence, M&S could have instructed Pilbrow and Partners to design a scheme that simply maximises commercial value*" (Fraser 2022a para 39, p. 11).

This view was backed up by architect Julia Barfield. She told the inspector:

> The brief here was clearly to maximise the site's potential and the architects have fulfilled their brief well – creating a building

minimizing operational carbon that, five or eight years ago, would have been considered fine. However, now that we understand the upfront impact of embodied carbon, it really isn't. Particularly building two extra basements! They are the worst in terms of embodied carbon.

(Hurst 2022)

M&S claimed that the environmental performance of the new build scheme would be of such a high standard that a number of years after completion it would be "a net positive contributor to the environment" due to low emissions required to run the building. Not only did we contest that these calculations did not include embodied energy from the construction required for the new scheme (40,000 tonnes of CO_2) but we raised concerns about the M&S data to support these claims.

Dr Julie Godefroy's evidence focused on M&S' whole life carbon assessment of their proposals and the sustainability data provided. She cast serious doubt on the pledge that the new building would deliver a net positive contribution to the environment from 17 years after completion (a figure M&S later changed to 11 years) (Marks and Spencer 2022). She stated in her evidence that the claim of a "payback" period would be much longer, most likely over 30 years, or it may not even be achieved within the standard 60-year assessment period. Godefroy added that as this claim was repeatedly presented by the applicant, and was used to justify the original decision by Westminster City Council to grant approval, adding, "*it is then of material importance that the claim is, in my opinion, not valid*" (Godefroy 2022 para 3.2.3.2, p. 5).

On the question of refurbishment options for the buildings, Matthew Fraser pointed out that in the face of a climate emergency, it is hard to justify such significant carbon emissions, particularly when it could be avoided if M&S seriously considered a deep refurbishment option.

Despite claiming that sustainability is at the core of their brand and committing to being a net zero business by 2040, M&S have dismissed the creative refurbishment alternative to such an extent that they have made a threat to leave Orchard House altogether if they do not get their way,

Mr Fraser told the inquiry. "*This is not the constructive attitude of a retailer dedicated to sustainability, heritage conservation and the future success of Oxford Street*" (Fraser 2022b para 32, p. 8).

He added:

One can sympathise with M&S for wanting a brand new building and releasing the commercial value that this would bring. However, the decision to proceed with a new build – without considering a comprehensive retrofit option – was made in 2018/early 2019 and it has never been revisited. Since then, climate legislation and regional and local planning policy has caught up with the terrifying reality of the global climate emergency.

(Fraser 2022b para 15, p. 4)

As part of our submission we argued that the basic refurbishment concept was to retain the existing structures and façades of the three buildings, strip out all the non-structural elements that have accumulated over the years, and add in the extra floors as set out by Arup at the inquiry. This would result in considerably less embodied carbon emissions and make use of existing resources – a principle tenet of our argument was that M&S failed to explore a deep retrofit approach.

One of M&S' main arguments has been that the floor to ceiling heights, column spacing and differing floor levels at some levels between the buildings meant that retention of the buildings is not practical. Throughout our campaign and evidence, we demonstrated that a growing number of refurbishment projects involving similar scale buildings of different periods shows that this type of refurbishment project is viable and possible. We also argued that M&S and their architects never fully tested comprehensive refurbishment options against the demolition scheme.

RA Lecture

While we waited for the result of the inquiry, we invited Simon Sturgis to deliver our annual lecture in March 2023 at the Royal Academy in London which he entitled: *"Architecture and the Climate Crisis: How the past can save the future"*.

Mr Sturgis used the event to call for a revolution in architectural thinking. He proposed a form of the Hippocratic Oath to "do no harm" to the environment, a provocation that made headlines in the architecture press. *"We also need imagination"* he declared. He went on to say

We need people to show much more imagination whether its architects or developers, the GLA, whoever it is on a national level. We need to show much more imagination with construction and with the design of buildings. Emissions from construction and use of buildings are now a bigger existential threat than nuclear war, just a lot less obvious or immediate.

(Hurst 2023)

Public Inquiry Success and Legal Challenge

Michael Gove ruled in July 2023 that M&S should not demolish its famous Marble Arch building. The decision was seen as a landmark moment in the debate about resources and re-use of existing buildings.

On sustainability, the Secretary of State's ruling, which overturned the recommendation of the planning inspector, stated that "*there was no dispute that the proposals would demolish and remove structurally sound buildings for a new larger development or that redevelopment would involve much greater embodied carbon than refurbishment*" (Gove 2023 para 21). He found this approach to be contrary to policy in the National Policy Planning Framework (NPPF). He interpreted the NPPF as reflecting a "*strong presumption in favour of repurposing and reusing existing building*" (Gove 2023 para 24).

The Secretary of State also did not consider there had been "*an appropriately thorough exploration of the alternatives to demolition*" (Gove 2023 para 32). Nor did he consider that M&S had demonstrated that refurbishment would not be deliverable or viable or that they had satisfied him that other options for retaining the buildings had been fully explored.

On heritage, the Secretary of State concluded that although Orchard House "*did not meet the listing criteria at the time it was considered for listing in 2021, it has significant value in its own right and in its context*" (Gove 2023 para 35) and he attached substantial weight to its loss. This was highly significant for an unlisted building, outside a conservation area. He found that the proposals conflicted with the NPPF and development plan policies on conservation of heritage assets.

High Court Challenge

M&S challenged the Secretary of State's decision on procedural grounds in relation to how he interpreted planning policy in reaching his decision, and how he explained his disagreement with the Inspector, following the public inquiry. The challenge was heard in the High Court in February 2024.

In her ruling on 1st March 2024, Mrs Justice Lieven allowed five of M&S' six grounds but upheld Mr Gove's heritage reasons for rejecting the planning application (Lieven 2024 paras 55, 63, 118, 132). The judge found that the Secretary of State had:

- misinterpreted the NPPF paragraph 152 as creating a "strong presumption" in favour of retaining existing buildings,
- failed to give legally adequate reasons disagreeing with his inspector's conclusions that there was no viable and deliverable alternative to the redevelopment scheme or the impacts for refusing planning permission, and
- made an error of fact in respect of the embodied carbon, and misapplied policy in this respect. The judge stated that he had mistakenly thought that London Plan carbon offsetting requirements applied to the embodied carbon of the existing building, not just the operational carbon of the redevelopment.
- The Secretary of State's heritage assessment was found to be lawfully reasoned.

Re-determination

Following the High Court ruling, the proposals returned to the new Secretary of State for re-determination. To inform this process, both sides were invited to submit representations between April and October 2024.

We continue to argue that M&S failed to comply with London Plan Guidance on the circular economy and whole life carbon assessments, by failing to sufficiently demonstrate that their buildings could not be viably and comprehensively retrofitted, thus avoiding the heritage destruction and embodied carbon footprint involved in the construction of the new building.

The new Secretary of State can approve or reject M&S' plans on heritage and/or sustainability grounds. There is nothing in the judgement which would stop her finding well-founded reasons for refusing the scheme on sustainability grounds, alongside the strong heritage reasons that were upheld. That decision is still pending. This is by no means the end of the story in this long running and high-profile case.

Re:store Competition

Figure 9.3 SAVE/*Architects' Journal* **Re:store ideas competition, Ravensbourne University.** (Ernest Simons/*Architects' Journal*)

**Figure 9.4 Team of architects taking part in design charette as part of SAVE/*Architects' Journal*
Re:store ideas competition, Ravensbourne University.**
(Ernest Simons/*Architects' Journal*)

In April 2024, SAVE and the *Architects' Journal* organised and hosted
Re:store, an ideas architecture competition intended to spark fresh
thinking about the M&S buildings and their capacity for re-use – see
Figures 9.3 and 9.4.

We were thrilled by the response from architectural practices, individual architects and students from across England. The quality of
entries was superb, with teams offering creative and technical
approaches to repurposing and reusing the buildings. A range of different uses were proposed to enhance the offer on Oxford Street and
many found ways of introducing natural light and "wow" moments
into their proposals as well as improving the public realm through and
around the buildings.

A long list of 13 teams was chosen to present their concepts to the
judges (made up of architects, academics and architectural historians)
who then picked six finalists. The finalists took part in a day-long charette workshop hosted by the school of architecture at Ravensbourne
University in North Greenwich. All their work featured in a special issue
of the *Architects' Journal* and in *Time Out* magazine. The competition
was also written up and published by the *Guardian*'s architecture critic
Olly Wainwright.[4]

4 *The Guardian*, 7 June 2024
https://www.theguardian.
com/artanddesign/article/
2024/jun/07/marks-and-
spencer-flagship-store-
london-art-deco.

Conclusion

This case has moved the debate forward by recognising that re-using buildings and treading carefully on the planet is a win-win scenario. It has captured the public's imagination and attracted widespread media interest and put carbon firmly at the heart of the debate. The campaign has challenged the UK's disposable attitude to buildings and was the first time a planning inquiry had sustainability and heritage as its joint focus. We have focused widespread attention on the wasteful knock-it-down and build again process that has dominated our construction sector for the last 100 years.

Now we urgently need robust national planning policy on retrofit that aligns with the UK Government's law on net zero targets. There is still no national policy requiring the assessment or control of embodied carbon emissions from buildings. The best way to remedy the lack of clarity is for the Government to update the NPPF to bring it up to date with the imperatives of the climate crisis. We also urgently need to regularise VAT rules between new build and existing buildings, and to close gaping planning loopholes around permitted development rights for the demolition of unlisted buildings outside conservation areas.

We need cultural change in the sector to challenge the idea that the "flight to quality" means a brand new sealed white box in a commercial building. We need to respond to the demands of the Gen Y and Z who are seeking employers and working space that respond to their values – characterful buildings, re-used spaces with high environmental credentials.

The current government was elected partly on its commitment to climate action and creating a zero-waste economy. Westminster City Council is in the process of adopting a proposed new "retrofit-first" policy, while recent research by consultants Lichfield's points to at least eight other London boroughs bringing forward similar policies including Camden and the City of London.

The dial is changing. Attitudes across the built environment industry are moving fast in favour of prioritising retrofit for climate, resource and heritage advantages. Allowing M&S to proceed with its wasteful and polluting plan is now seen as indefensible.

There is Another Way

Agile and forward-thinking property developers working in central London tell us that their tenants are increasingly demanding characterful commercial retrofits that "earn the commute", not the outmoded glass and steel boxes, which come with a vast embodied carbon footprint attached.

A more sustainable trend for re-using department store buildings on Oxford Street and elsewhere is emerging. Comprehensive retrofits such as the 1930s former DH Evans/House of Fraser building also on Oxford Street will see the historic façades restored and retail uses topped with offices and a rooftop restaurant. The former Debenhams flagship store will reopen next year with new office space and roof terrace and retail, and the former Top Shop at Oxford Circus soon to be a town centre IKEA are all in the pipeline. Nearby the former Fenwicks store on New Bond Street is being repurposed by Foster and Partners as retail on the lower floors with office space above.

Outside London, Bobby's department store in Bournemouth has been brought back to use as a multi-use space including shopping, dining and events. There are plans to convert a former Debenhams in Harrogate into flats over two floors of flexible commercial space. Primark has taken on the famous Bank Buildings in a heroic restoration project in Belfast. In Liverpool, the 1925 Owen Owen department store has just been restored by Flannels – boasting seven floors of retail, a fitness studio and four bars and restaurants.

All these examples demonstrate a more resource- and carbon-efficient approach, compatible with the UK's trajectory to net zero.

Reference List

Bennet, C. (2022). Razing your architectural gem is a funny way to show a love for heritage, M&S. *The Guardian*, 29.10. 2022. https://www.theguardian.com/commentisfree/2022/oct/29/razing-your-flagship-building-is-a-funny-way-to-show-a-love-for-heritage-marks-and-spencer

Climate Change Committee (2023). *Progress in reducing emissions: 2023 Report to Parliament*, p. 33. https://www.theccc.org.uk/wp-content/uploads/2023/06/Progress-in-reducing-UK-emissions-2023-Report-to-Parliament-1.pdf

Elefante, C. (2007). The greenest building is ... one that is already built. *Forum Journal*, 21(4), 26–38.

Environmental Audit Committee 2022-23. *Building to net zero: Costing carbon in construction First Report of Session 2022–23* edited by House of Commons. London: UK Parliament.

Forshaw, A. (2022). *Proof of evidence on behalf of SAVE Britain's Heritage.* 27.9.2022. www.savebritainsheritage.org/docs/genera/AF_PoE_Final.pdf

Fraser, M. (2022a). *Closing Submissions by SAVE Britain's Heritage MARKS & SPENCER, 456-472 OXFORD STREET, LONDON PCU/RTI/X5990/3296903.* Department for Levelling Up, Housing & Communities. 4.11.2022. https://www.savebritainsheritage.org/docs/articles/Oxford_Street_MS_SAVE_Closing_FINAL.pdf

Fraser, M. (2022b). *Opening Statement at the public inquiry by Matthew Fraser of SAVE Britain's Heritage Marks & Spencer, 456-472 Oxford Street, London PCU/RTI/X5990/3296903.* Department for Levelling Up, Housing & Communities. 25.10.2022. https://www.savebritainsheritage.org/docs/articles/SAVE_Opening_Statement_FINAL_-_MF.pdf

Godefroy, J. (2022). *Proof of Evidence by Dr Julie Godefroy on Behalf of Save Britain's Heritage.* Department for Levelling Up, Housing & Communities. . 27.9.2022. https://www.savebritainsheritage.org/docs/general/JG_PoE_Final.pdf

Gove, M. (2023). Town and Country Planning ACT 1990 – Section 77 Application Made By Marks and Spencer PLC 456-472 Oxford Street, London W1 Application REF: 21/04502/FULL. Edited by Housing & Communities Department for Levelling Up. London: UK Government.

HOC (2024). *The UK's plans and progress to reach net zero by 2050, Research Briefing, para 1.1.* London: House of Commons Library.

Hurst, W. (2022). From Kristin Scott Thomas to Julia Barfield: Who is saying what about M&S Oxford Street. *Architects' Journal.* https://www.architects journal.co.uk/news/from-kristin-scott-thomas-to-julia-barfield-who-is-saying-what-about-ms-oxford-street

Hurst, W. (2023). Architects need a Hippocratic oath to stop them trashing the planet. *Architects' Journal.* https://www.architectsjournal.co.uk/news/architects-need-a-hippocratic-oath-to-stop-them-trashing-the-planet

Jenkins, S. (2022). M&S is a shining example of how not to treat the high street – or the planet. *The Guardian*, 28.10.22. https://www.theguardian.com/commentisfree/2022/oct/28/m-and-s-high-street-retailer-flagship-store-small-shops

Lieven, Justice. (2024). Between: Marks and Spencer Plc Claimant and Secretary of State for Levelling Up, Housing and Communities First Defendant and Westminster City Council, Second Defendant and SAVE Britain's Heritage Third Defendant 1st March 2024. In *Neutral Citation Number: [2024] EWHC 452 (Admin) Case No: AC-2023-LON-002520*, edited by The High Court of Justice King's Bench Division Administrative Court Planning Court. London: Judiciary UK.

Lloyd, H. (2022) *Departing stores: Emporia at risk.* Save Britain's Heritage.

Ludik, M. (2022). Plans to demolish M&S Oxford Street store puts historic character of London at risk. *The Evening Standard*, 30.10.2022. https://www.standard.co.uk/business/marks-and-spencer-oxford-street-store-london-michael-gove-b1036397.html

Marks and Spencer (2022). Bold, sustainable and innovative: M&S sets out comprehensive case to deliver fit for the future Marble Arch, https://corporate.marksandspencer.com/media/press-releases/bold-sustainable-and-innovative-ms-sets-out-comprehensive-case-deliver-fit

Mayor of London (2021). *The London Plan.*

Royal Institution of Chartered Surveyors (RICS) (2023). *Whole life carbon assessment for the built environment* 2nd Edition. London: RICS.

Saint, A. (2020). *Oxford Street.* Vol 53. Survey of London. London: Paul Mellon Centre for Studies in British Art.

SAVE Britain's Heritage (2021). SAVE urges re-think on historic M&S flagship store demolition plans. 23.11.21. https://www.savebritainsheritage.org/campaigns/article/754/save-urges-re-think-on-historic-ms-flagship-store-demolition-plans

SAVE Britain's Heritage (2022a) Leading figures urge Communities Secretary to hold inquiry into M&S demolition plans. 18.5.2022. https://www.savebritainsheritage.org/campaigns/article/791/press-release-leading-figures-urge-communities-secretary-to-hold-inquiry-into-ms-demolition-plans

SAVE Britain's Heritage (2022b). Letter to Secretary of State with signatories. 18.5.22.

SAVE Britain's Heritage (2022c). New report blasts bulldoze and rebuild plan for M&S Oxford Street HQ. 21.1.2022. https://www.savebritainsheritage.org/campaigns/article/766/press-release-new-report-blasts-bulldoze-and-rebuild-plan-for-ms-oxford-street-hq

Sturgis, S. (2022). *Report by Simon Sturgis, published January 2022, commis-sioned by SAVE Britain's Heritage.* https://mcusercontent.com/9a03bb1
1e3ccc82634488e2b/files/07cd2fe4-f3df-3cdf-ba83-5d119660f367/220120
FINALSSM_S.pdf
Waite, R., Highfield, A., & Hurst, W. (2023). BREAKING: Gove rejects M&S
Oxford Street demolition. *Architects' Journal* https://www.architectsjournal.
co.uk/news/breaking-gove-rejects-ms-oxford-street-demolition

10

CONCLUSION

Oriel Prizeman

We wait for the short-term interests of democratic governments globally to tentatively contemplate the role of embodied carbon in planning legislation. It is evident that that time for the question of long-term responsibility to future generations to be adopted and recognised as a matter of course, must be a cultural as well as a regulatory shift. In Chapter 2, Peter Cox established the importance of whole life cycle analysis as opposed to operational energy as a necessity for the consideration of the construction industry as a whole. In Chapter 3, Deepankar Kumar Ashish and Riccardo Maddalena both outline the significantly increased degrees of embodied carbon of relatively modern building materials, specifically concrete and steel. They also point to future means to reduce these levels in the development of alternative, lower carbon aggregates etc. Jigna Desai and Rajan Rawal (Chapter 4) highlight the impact in India of the legacy of significant numbers of post-independence reinforced concrete and modern buildings. In a detailed analysis of their own campus building and its sympathetic passive design strategies, it provides a positive indication of the potential for such analysis to support the conservation of modern heritage.

With the exception of the examples in Dublin presented in Chapter 8, the overall focus of the contributions here emphasises how the oppositional imperatives of the last century in terms of energy use and construction techniques have become a legacy for this one. In particular, the significance of concrete as a carbon intensive material is drawn out in Chapter 2 with regard to materials science but also in the analysis of modern heritage in India in Chapter 3. The key themes that emerge might be summarised as: 1. a change in values; 2. the debt to cement and steel; 3. the importance of maintenance, management and husbandry; and finally, 4. the new powers of persuasion. Overall, as anticipated in the introduction, the issue of embodied carbon broadens the focus and remit of a sustainable approach to building conservation to accommodate not just the adaptive reuse or careful modification of heritage buildings but to adopt a conservative attitude to extending the life of much more marginal buildings. The findings are summarised below.

DOI: 10.4324/9781003527404-13

A Change in Values

At a global level, UNESCO's operational guidelines define Outstanding Universal Value as

> cultural and/or natural significance, which is so exceptional as to transcend national boundaries and to be of common importance for present and future generations of all humanity. As such, the permanent protection of this heritage is of the highest importance to the international community as a whole.
>
> (UNESCO 2023)

Historic England's enduring English Heritage guidance document "Conservation Principles" outlines a means to approach the definition of "significance" through the attribution of four measures of value: Evidential, Historical, Aesthetic and Communal (English Heritage 2008). The adoption of a duty of care with regard to embodied carbon before deciding to demolish a building will typically not apply to buildings that already enjoy statutory protection derived through the application of such values.

The marriage of sustainability and conservation principles must become more critical. As outlined by Peter Cox in Chapters 2 and 8, new thinking is required to perhaps add embodied carbon as a criterion for consideration. Overall, it is evident from the majority of examples considered here (the apartment buildings in Alexandria in Chapter 7, most of the UK Carnegie libraries in Chapter 5, the M&S building in Chapter 9) that the critical cases are those at the margins, not the heart of our heritage protection systems.

The Debt to Cement and Steel

Traditional buildings globally – commonly referred to as pre-1919 or pre-20th century, tend to be constructed using natural materials: Lime or earth-based mortars for masonry as well as for plastering and rendering timber framed structures using local knowledge. Characterised as breathable and even marginally flexible materials, their performance is well understood to be completely opposite in performance terms to the impenetrable, impregnable technologies of building materials since the start of the 20th century, also globally. The use of steel framed construction has transformed the size and scale of human inhabitation. The great facility, cheapness, malleability, strength and instantaneous capacity of concrete to allow humans to build at scale has transformed the planet in the last century, as anticipated by Le Corbusier with his focus on the future of standardisation.

However, the hugely carbon intensive process of cement production and thus concrete construction as well as that of steel adds a significant debt in terms of embodied carbon to all structures made with these materials. This rationale explained in detail from the perspective of materials science in Chapter 3 and illustrated vividly in the life cycle case study of concrete buildings in Ahmedabad, India in Chapter 4. The quantity and thus relevance of associated similar potential examples globally is inestimable.

The Importance of Maintenance, Management and Husbandry

UNESCO's operational guidelines call for the protection and management of world heritage sites. Taking this duty as a remit that could be applied in places of lesser significance is a fundamental lesson of considering embodied earbon. In Chapter 6, the management of processes of grassland burning in the Aso cultural landscape of Japan provides an exemplar of stewardship. Management, however, is also about the intelligent use of buildings. In Chapter 5, the review of actual energy use in over 400 turn-of-the-20th-century public library buildings in the UK identifies how frugal management and original design intentions contradict predicted assumptions of poor heat energy performance and electricity use. The potential for passive ventilation strategies to be re-introduced to a predominantly air-conditioned apartment building in hot humid Alexandria is outlined in Chapter 7. The modelling illustrates how an adaptive approach to behaviour could also protect and enhance the future of a marginalised and abused late 19th-century building.

New Powers of Persuasion

Numerous failed campaigns to "list" significant 20th-century heritage in the UK including John Madin's 1975 colossal concrete Birmingham library have sought and failed to gain heritage protection as a means to avoid demolition. These appeals were based predominantly on the elusive and subjective definitions of architectural merit – a theme that is sadly seldom of interest to Secretaries of State. There is evidence, however, as outlined in Henrietta Billings' account of the M&S case, thanks to the quantifiable aspect of embodied carbon, that politicians are beginning to be able to consider sustainability in such cases as a relevant factor in the public interest. The increased power of persuasion enabled by using a quantifiable as opposed to a qualitative metric for traffic light decision makers is revolutionary. This precedent of this campaign potentially changes everything.

However, although there was a consultation in September 2024 on the presumption of considering retrofit before demolition as a planning policy in England, the current consultation paper for National planning policy makes 73 mentions of the word "existing", not one refers to a building. The 2023 United Nations Progress chart for the Sustainable Development Goals highlights that for goal 11, "Sustainable Cities and Communities", there is more insufficient data than for any of the other SDGs. Indeed, these insufficient data include those for 11.c "Sustainable and Resilient Buildings" (United Nations 2023). It is surely possible for the mainstream adoption of accounting for embodied carbon to both provide these data and support better decisions in support of this goal.

Reference List

English Heritage. (2008). *Conservation principles, policies and guidance.* https://historicengland.org.uk/images-books/publications/conservation-principles-sustainable-management-historic-environment/.

UNESCO (2023). *Operational guidelines for the implementation of the World Heritage Convention,* edited by Scientific and Cultural Organization United Nations Educational, Intergovernmental Committee for The Protection of the World Cultural And Natural Heritage. Paris: World Heritage Centre.

United Nations (2023). *The Sustainable Development Goals progress chart.* https://unstats.un.org/sdgs/report/2023.progress-chart/Progress-Chart-2023.pdf.

Index

Numbers in *italics* refer to text within *figures*. Numbers in **bold** refer to text within **tables**.